THE ASPIRING FIREFIGHTER'S TWO-YEAR PLAN

by
Paul S. Lepore

The Aspiring Firefighter's Two-Year Plan

The Complete Road Map to Becoming a Firefighter

ISBN: 978-0-9729934-3-2

www.aspiringfirefighters.com

Printing History:
September 2004 First edition
April 2007 Second edition

Printed in the United States of America

Dedication

This book is dedicated to the men and women in the fire service, who are willing to give up their life for a complete stranger.

Paul Lepore is a Battalion Chief with the City of Long Beach, California, Fire Department. He entered the fire service as a civilian Paramedic for the Los Angeles City Fire Department in 1985. After completing his education at the Los Angeles Paramedic Training Institute, Lepore worked in the high impact area of South Central, Los Angeles.

Lepore was hired by the Long Beach Fire Department in 1986. He spent two years working as a firefighter, nine years as a firefighter/paramedic, seven years as a fire captain, and is currently a Battalion Chief.

He has conducted hundreds of entry-level interviews as well as served as a rater for Fire Captain and Battalion Chief promotional exams. He holds instructor credentials for EMT, Hazardous Materials, and Weapons of Mass Destruction.

Lepore has conducted numerous seminars to coach and mentor both promotional and entry-level candidates alike. Lepore is the author of <u>Smoke Your Firefighter Interview</u> and <u>Smoke Your Firefighter Written Exam</u>, which help candidates achieve their goal of becoming a firefighter.

He founded EMS Safety Services, Inc., a first aid and CPR training corporation that certifies over 100,000 students and instructors annually.

He is an avid saltwater fisherman and has written a fishing book entitled, <u>Sport Fishing in Baja</u>.

Paul Lepore and his wife Marian have two daughters, Ashley and Samantha. They live in Dana Point, California.

www.aspiringfirefighters.com

800.215.9555

Table of Contents

Two-Year Plan...1

What it Takes to Land a Job 9

Extenuating Circumstances...13

Preparation

  The Importance of Education..21

  Station Visits 27

  Ride-Alongs..33

  Role Model..39

  Fire Technology Courses and the Basic Fire Academy........45

  Friends and Acquaintances..59

  Private Interview Coaching 63

Hiring Process

  Written Examinations 67

  Interviews...73

  Situational Questions 81

  The Chief's Interview 87

  Resumes..91

  Physical Fitness 99

  The Physical Ability Test..107

  Background Investigations 121

  Polygraph Exams 135

Personal Attributes

  Public Speaking 141

Table
of
Contents

Mechanical Ability 147

Community Service 153

Tattoos.. 159

Age 163

Military Experience.. 167

What to Expect

The First Day... 173

Rookie Life 177

Fire Academy 185

Paramedic School 191

Station Drills.. 199

Dangers of the Job.. 215

Other Occupations

Should I Become a Paramedic?....................................... 221

Becoming a Police Officer 229

Driving an Ambulance 233

Different Perspectives

Female Firefighters 239

Firefighting: A Wife's Perspective 243

Having a Firefighter for a Parent 253

God's Cake.. 255

Appendices

Interview Grading Sheet.. 259

Probationary Firefighter Evaluation................................. 263

Two-Year Plan

This sample Firefighter Two-Year Plan was developed with input from Mike Sarjeant, a Battalion Chief on the Long Beach, California, Fire Department.

- **If still in high school look into a Regional Occupational Program (ROP).**

  Many local fire departments have community outreach recruitment programs.

- **Graduate from high school or obtain your GED.**

  A diploma is much preferred.

- **Talk with a counselor at a community college that offers fire science courses.**

  Set up a course curriculum that allows you to obtain a two-year degree in fire science. If the local college does not offer a fire science program, find one that does.

  This curriculum should also allow you to complete the prerequisite courses for a fire academy.

- **Take an Emergency Medical Technician Course (EMT).**

  This will accomplish a few things. First of all, it is a course required by most departments. It will also let you know if this profession is for you. If you find you can't handle the sight of blood or helping people during crises, the fire service may not be for you.

- **Enroll in a state certified fire academy.**

  Many departments require completion of a Firefighter 1 Academy prior to taking the entry-level exam.

  Completion of a fire academy prior to being hired will greatly enhance a candidate's chance of successfully completing the fire department's academy. Many fire departments have a 25-30% failure rate.

- **Find out if your community has either a fire department volunteer program or Fire Explorers.**

  Volunteering in the fire department is an excellent way to gain real life experience. This exposure will also allow you to determine if this is indeed the right career choice for you.

- **Volunteer in your community.**

  Find something that you are interested in and volunteer your time: church, sports, hospital, YMCA, Red Cross, etc. It doesn't matter. Get involved. Volunteering is something that should be done because it's the right thing to do, not because it will look good on a firefighter application.

  Firefighters are self-motivated and have historically been involved in their communities. The perception is that if you are helping out in your community now, you will be the type who will likely continue to stay involved after you are hired, helping out in various committees and groups both on and off the job.

- **Visit the local fire stations.**

  Interview the firefighters and elicit their help in planning your career path. It is a tremendous compliment to the firefighters to have someone aspire to be in their position. Visiting the fire stations will help you learn about the job and the culture of the fire service. In addition, you will learn of things that you could be doing to enhance your chances of getting hired. Ultimately, when the department hires, you will be in a good position since the firefighters have gotten to know you and have taken the time to mentor you. There is nothing better than a "home grown" prospect.

- **Prepare for a fire department interview.**

  Consider the reasons why you want to become a firefighter and be able to express them. Do your research and learn the rules of the road concerning the interview process. Participate in "mock" interviews with firefighters.

- **Start a log that includes everything you have done to prepare yourself.**

  Include details, dates and names of instructors. Include any personal experiences that may be pertinent to becoming a firefighter.

  A few examples of this could be:

  > You witnessed a car accident and were able to render aid.

  > You volunteered your time at the Boys and Girls Club.

  > You experienced a life-changing event.

  > You were voted most inspirational on your athletic team or your fire academy.

  > Your high school athletic team won the championship.

  > You were a lifeguard at the city pool.

  > Anything that you think might be significant. There are no rules. Write it down!

  This information will go on your resume, or may be speaking points in an interview. This is preparing you to answer difficult questions in an interview, such as, "Please share with the panel a stressful time in your life and how you dealt with it."

  The log should just be an easy and accessible memory jogger for you. If you are comfortable with a pencil and notepad, keep them in your room in a convenient spot so you won't forget to use them. If you are more comfortable on the computer, then use it to formulate your thoughts and ideas.

- **Get in shape.**

  Firefighting is a very physical job requiring peak physical strength and endurance. If you are not in good physical condition, it will become very evident during the physical ability testing or the pre-hire medical exam. It is also important to look as if you are physically prepared for the job.

  If you see a firefighter who looks out of shape, don't look at him and think, "If he got hired, so can I!" Odds are he was in better physical condition when he was first hired. You are trying to do everything you can to improve your chances. This is a very important part that you have complete control over.

- **Look the part!**

  The rule of thumb in an interview is to hire someone who you can see becoming a member of your crew tomorrow. A candidate who walks in with excessive facial hair, large tattoos or body piercing that is not permitted by the department's policies presents as a candidate who is not ready for the position. Do not make the mistake of saying that you will remove them when you are ready to be hired. You are making a statement. It is important to understand that the fire department is a paramilitary organization. These will definitely not improve your chances of success.

- **Dress professionally.**

  Invest in a suit and tie. Although not required for the interview, a candidate who does not wear one stands out. First impressions are critical. Make sure the suit is conservative, not flashy.

  Dress professionally whenever you will have contact with members of the department. This includes station visits. Remember, it is important to make a good first impression.

- **Enroll in a service that lets you know which departments are testing.**

  There are several businesses on the Internet that will inform you of which departments are testing and what their requirements are.

  Most departments test every two to three years. They will then hire from the "eligibility list" until it expires. The window to file an application is usually very small, ranging from as short as one day to as long as 30 days. Once the filing period is closed, the department will not accept any more applications. If you don't have a subscription to one of these services, you will miss a lot of opportunities.

- **Talk to your family.**

  The decision to become a firefighter is a monumental one. It will most likely be a long road that requires a lot of time and sacrifice. If you don't have a family or friend support network, it will become extremely difficult. Most importantly, if your spouse does not support your decision, you are destined for failure.

- **Surround yourself with reputable people.**

  A firefighter position is a life choice, not just a job. You must be prepared to live your life with excellent moral and ethical values. For this you will need the support of family and friends who are good role models. If your friends are not a positive influence in the community, you may want to find a new set of friends. Remember the old saying, "Birds of a feather flock together." A background check will scrutinize not only you, but also the company you keep.

- **Learn a trade.**

  Woodworking, framing, electrical, plumbing, welding and automotive are all common examples of a trade. Firefighting is a very physical job that requires good psychomotor skills and a hands-on approach. Typically those who have learned a trade possess these applicable job skills. If you know how a building is constructed, you will be able to predict how a fire will travel through it. If you know where the electrical and plumbing is typically run behind the drywall, you will most likely know where it would be safe to open it up. You will also have become very comfortable with power tools. The importance of being able to work with your hands cannot be overstated.

  If you don't currently have this kind of experience, start taking classes in a trade at your community college. You will at least learn the basics. Back this up with some real life practical experience. It will be invaluable knowledge and will play out well in an interview. Mechanical aptitude cannot be learned in an Internet class or while sitting behind a computer.

- **Improve your public speaking skills.**

  If you are uncomfortable getting up in front of a group, you must take steps to overcome your fear. The largest percentage of the testing process is the interview and ultimately a large part of the job deals with public speaking! You won't talk a fire out, but you will talk to different groups about how to prevent them. If you can present yourself well in an interview, you are leaps and bounds ahead of the others who can't. Even if the other candidates have more experience than you, the job

will usually be awarded to the candidate who can present him or herself in a clear and concise manner.

If public speaking is your downfall, it is imperative to join Toastmasters or take some courses at your community college. A speech and debate class is an excellent way to get over the jitters. Acting or drama classes can also be an excellent way to feel more comfortable in front of a group.

Teaching others can also help you learn to think on your feet. Whether you are teaching CPR and First Aid or your local Sunday school class, it will help you learn to present information clearly and field questions.

A typical interview question might be, "What do you consider a weakness about yourself?" Your answer could be, "I used to feel uncomfortable getting up and speaking in front of a group. I knew this was a very important part of my chosen vocation. I took several classes at my community college to help improve my comfort level. Since then I feel much more confident in my ability to speak in public."

You can have all of the best traits in the world, but if you can't effectively convey them in an interview they will go unnoticed. Now that's turning a negative into a positive!

- **Maintain a clean driving and criminal record.**

  It goes without saying that firefighters are held to a standard that is much higher than the average citizen. The road is littered with firefighter candidates who have failed their background check due to a poor driving or criminal record.

- **Maintain a good credit history.**

  Your credit history is a reflection of your reliability, honesty, organization and attention to detail.

- **Update your resume.**

  Make sure your resume has no technical or grammatical errors, is well organized and comprehensive. Ask reliable friends or family to proofread it.

Notes:

A wise man knows what to say.
A wiser man knows when not to say anything.

Firefighting is a challenging and rewarding career.

What it Takes to Land a Job

Landing a job in the fire service is truly a unique challenge. On average, there are over 100 candidates who apply for each opening. Since the competition is so intense, what does it take to be the top candidate?

Many candidates believe it is important to be the "most qualified" individual in the testing process. The truth is that we are looking for someone who will fit into our family. In short, we have an opening that we need to fill. Since we can choose whomever we want, we want to choose someone we like. Those candidates who become known to us either before or during the testing process have a better chance of scoring well on the exam.

The best way to become someone who stands out in the hiring process is to understand the role of a firefighter. This can best be accomplished by taking fire science courses at the local junior college or online. Another way to gain knowledge and experience in the fire service is to become a volunteer or

reserve firefighter. These candidates will have made a name for themselves long before the testing process.

Candidates often volunteer for departmental activities. These activities include departmental BBQ's, CPR training events for the community and any other opportunities that may arise to give a candidate a chance to be visible to the members of the department. As you are flipping burgers, it is entirely possible that a captain, battalion chief or even the fire chief will stop you and introduce him or herself. This is your opportunity to meet influential people on the department. Once the introductions are made, the conversation often steers toward what you are doing. This is your opportunity to explain that it is your goal to become a member of the department.

Most departments have a minimum passing score for the written exam and physical ability tests. This leaves the bulk of the score (oftentimes 100%) for the oral interview. Since we are looking to hire people we like and want to have as part of our family, it is

> **The best way to become someone who stands out in the hiring process is to understand the role of a firefighter.**

imperative that the oral board knows who you are before you walk in the door. This may be extremely difficult on a large department since there are just too many people to meet. On a smaller department it is possible to "make the rounds" to all or most of the fire stations before your oral interview. Imagine what an incredible opportunity it would be to take a practice interview with experienced firefighters.

It is important to note that you are establishing your reputation the minute you walk into the fire station. If you make a favorable impression, the firefighters will help you and maybe even pass positive information to the oral board. The same thing can be said if you make a poor showing.

It is impossible for the board to get to know you within a 20–30 minute interview. A candidate who maximizes his or her time before the interview by spending time in the stations and getting to know the firefighters can vastly improve his or her score. If the firefighters like you, they can put in a word to the oral board. If the oral board doesn't have a good feel for you there is no way you will score in the top.

The way we score candidates is different than most people would expect.

If the board consists of two or three firefighters, the minute you walk out the door we look at each other and try to decide if we want you on our crew. If the interviewers really like you they will score you in the high 90's. If they thought you were the average "vanilla" candidate with the usual complement of fire science classes, maybe even the academy and a reserve firefighter position, you will be in the low to mid 80's. If the board really doesn't like your demeanor or feels like you were completely unprepared for the interview, you will be below the minimum score of 70%.

If you have already taken fire department examinations, reflect back to your oral interview scores and try to interpret what the board is trying to tell you. If you are in the high 90's I would suggest that you make sure you are in top physical condition. You are on the brink of being hired. Don't change what you are doing, as you are already on the right track.

If you have taken a plethora of fire science courses and are doing all the right things to get a job, but still find yourself in the low to mid 80's, you need to re-evaluate how you are performing on your interviews. It is important to remember that it is not about having more qualifications than the next candidate; it's about coming across as someone we want to have on our crew. If you already have all of the wallpaper (certificates and classes) and you are not scoring well, you have a serious problem. It's time to seek some outside advice. Your best bet is to find as many people as you can to give you mock interviews. Hopefully someone can identify what you are doing wrong and stop you from spinning your wheels.

If you scored below a 70% and this was your first or second exam, don't worry, as it's a long process. Continue taking fire science courses and learn as much as you can about the fire service. The more you understand about our culture and idiosyncrasies, the more you will be able to prove you are ready for the position.

The fire service is a unique occupation. There is no matrix to follow to ensure you will be offered a position. It's actually the opposite. A candidate can have all of the horsepower known to mankind and still not be offered a position, while a candidate who has never taken a single class is offered a job on his or her first examination. To an outsider it may be quite perplexing. To an insider we all understand it is about being the person we all want to have on our crew. It really is not that complicated.

Notes:

Extenuating Circumstances

Everybody has made mistakes in life, and if confronted with the circumstances a second time, would do things differently. I believe they call this "life experience." While the fire service is looking for people who exercise good judgment, we also realize that people make mistakes throughout the course of a lifetime. No one is perfect.

How grave a mistake will the fire department overlook is anyone's guess. I believe it also depends on the value system of the person making the ultimate decision, the fire chief. If he or she is a strict disciplinarian, the chances are less favorable that a candidate with a tarnished past will be hired.

Some people who have made mistakes in life recover and never look back. Others make a significant mistake and continue to make the same poor choices. A person with a history of errors in judgment does not stand much of a chance of convincing the fire chief that he or she has changed.

I have a friend who is the CEO of a small corporation. Recently, the company was hiring for a certain position. Due to the poor state of the economy, the phone had been ringing off the hook. The Operations Director narrowed the field to six candidates. The CEO sat in on the final interviews.

The Operations Director prioritized the candidates and the interviews began. His "favorite" candidate was on deck. As he reviewed her application prior to the interview, the CEO was impressed with her strong work history. As they read down the application they were surprised to see that she had been terminated from Bank of America.

The candidate entered the interview room and was dressed professionally. She fielded all of the interview questions with ease. Predictably, the interview led toward her prior position at Bank of America. She looked the CEO in the eye and said she had been terminated. He was taken aback at her forthrightness. He expected her to sugarcoat the situation so it would look as if she had been wrongfully terminated. He asked the applicant if she cared to elaborate.

The applicant explained that it was B of A policy to close the teller window after a large cash deposit. A merchant came in and made a large cash deposit.

Two-Year Plan

We are quick to criticize,
yet slow to recognize.

The applicant noticed that the line of customers stretched out the door and there was only one other teller on duty. She called for another teller but no one else came.

In the interest of customer service, she broke policy and elected to stay open. The next customer came in with another large cash deposit. As luck would have it, the third customer came in and held up the bank. Even though she was the lead teller (and customer service representative), she was terminated for breaking company policy.

The CEO imagined himself standing in line at the bank and seeing only one teller and he applauded her decision in the interest of customer service. If she had closed her window he could imagine himself being irate. It also concerned him to hire someone who would selectively follow the policies and procedures.

The applicant broke policy and was terminated (which is exactly what the CEO would have done), but he respected her decision nonetheless.

The applicant never tried to blame anyone else for her mistake. She took full responsibility. The CEO respected her for standing by her decision. As a result, she was hired over 75 other applicants.

If she had tried to rationalize her mistake, she would have never stood a chance or even made it to a second interview. The way she conducted herself earned her the job. It was not the mistake, but how she handled herself after the situation.

Another situation occurred with my friend Rob, who was the captain of a Sportfishing yacht at the young age of 21. He was years ahead of his time, being given the responsibility of skippering a 44-foot yacht based out of Newport Beach, California. His job was to take clients fishing 75 miles offshore. He quickly developed a reputation as a good fisherman and a capable captain.

Unfortunately, Rob's success got the best of him. He quickly became the "big man on campus." While running the yacht late one evening, Rob fell asleep behind the wheel. They were headed to one of the offshore islands. Since the autopilot was set to the middle of the island, the boat ran aground when it reached its destination. The rest of the crew and the passengers were thrown from their bunks onto the floor. Thank goodness everyone was okay.

Two-Year Plan

Stand up for what you believe in.

Fortunately, the boat ended up on one of the few beaches on the island. Hours later a salvage company was able to pull the boat from the beach and tow it back to the harbor. The damage was extensive and the boat was out of commission for several days. As you can imagine Rob's world came crashing to the ground. He was fired from his job as a captain and his future was uncertain.

Instead of burying his head in the sand, he stood up and apologized for his mistake. He took full responsibility for his actions. Did he have trouble finding another job? You betcha!

After a series of interviews, he earned a job making great money running a 1.5 million dollar yacht. When the owner of the yacht was asked if he was concerned about Rob falling asleep behind the wheel again, he replied, "Not a chance." Anyone who has ever experienced something of that magnitude would never want to go through that again. Who would you rather have running your yacht while you were asleep?

The fire service is full of everyday people who have made mistakes in life. Many of these individuals were given a second chance at pursuing their dreams. People have overcome problems with drugs, credit, driving under the influence and run-ins with the law. The intent of this chapter is not to determine what will disqualify you for a firefighter position, but rather to shed some light on the process.

Everyone has made mistakes in life. It is what you do to recover after them that demonstrates your true character.

Notes:

_____

_____

_____

_____

_____

Notes:

Preparation

The Importance of Education

On the surface it may seem that education is not important for a firefighter. This is very far from the truth for several reasons. First of all, firefighters have evolved from "put the wet stuff on the red stuff," to being in charge of major incidents involving hazardous materials or weapons of mass destruction, or determining paramedic level care on a gravely ill or injured patient. What do all of these incidents have in common? Each discipline requires knowledge of physics and chemistry, of course!

Secondly, firefighters are required to write a report (which is also a legal document) that summarizes every emergency response. These reports are a direct reflection of the report writer. If a report is filled with grammatical and punctuation errors, the credibility of the writer is brought into question. Firefighters are often asked to testify in a court of law as to what occurred. A firefighter who authors a report riddled with errors will certainly lose credibility with the audience.

Firefighters negotiate their salaries with the city, county or board. The more educated the firefighters are in the political process, the better they will fare at the bargaining table. This ultimately translates into better wages, benefits and working conditions.

Is a Bachelor's degree required prior to getting hired? The answer is no, or at least not in most places. Most departments require a high school diploma or a GED certificate. Why is there such a wide range of education levels for entry-level firefighters? It's really quite simple. The person making the hiring decisions sets the tone as to the importance of education. If the fire chief values education, you can bet he or she will expect the entry-level firefighters to have a degree (or at least be actively working toward one) prior to getting hired. If, on the other hand, he or she is more mechanically inclined, education may not be a priority. These organizational priorities change as the fire chief retires, and the new fire chief will set his or her own priorities.

I began taking my fire science courses shortly after having graduated from high school. I entered the fire science program at the local junior college, taking the six fire science and EMT prerequisite classes for the basic fire academy I completed the courses in two semesters and one summer session, then enterered the fire academy. Upon graduation from the fire academy and armed with 30 units of fire science courses, I started picking away at my Associate of Science degree in fire science. I was fortunate enough to be hired at 20 years old by the Los Angeles City Fire Department as a single function paramedic. Eighteen months later, I was hired by Long Beach as a firefighter.

I had great intentions of completing my Associate's degree and ultimately my Bachelor's degree. A promotion to firefighter/paramedic and ultimately to Captain, starting a business, becoming a husband, father and author has put my educational plans on hold. In short, the rigors of dealing with everyday life as a firefighter and the shifting schedule made it difficult to continue my education. Is this an excuse? No way. I firmly believe that anything can be accomplished once you set your mind to it.

Is it possible to get your education after getting hired on a fire department? By all means, yes. At age 39, I went back to school. I earned my Bachelor's degree while working as a Battalion Chief. If I can do this while working full time, and being a husband and father, so can you.

In today's day and age, the advent of the Internet makes it possible for a student to complete a course regardless of the time or location. There are numerous colleges which now offer fire science courses online. These are the perfect solution for a working person with a family who struggles to get into a structured class. The student does not have to worry about getting off work early, fighting traffic, paying for parking, or finding a babysitter for the kids. Online courses accommodate all schedules, since it does not matter what time of the day or night a student "logs in" to participate in the discussion centers. In my opinion, there is now no excuse for a person applying to fire departments not to have his or her education.

In many areas of the country an Associate's degree is the standard. If a candidate does not have one, the evaluator's eyebrows are raised to question why he or she has not taken the time to earn one. In a few communities it is even required before taking the entry-level exam.

Many new firefighters often have more advanced degrees. Although this depends on a myriad of different circumstances, it seems there is certainly a strong trend in this direction.

Where do these more highly educated candidates come from? Are these the same fire science students found in the average fire science courses? No, commonly they are people who obtained a degree to enter the professional workforce as a teacher, computer specialist, stockbroker, or some other profession, but decided they were dissatisfied in their profession. In short, they decided on a career change.

As a general rule these candidates are older than the typical applicant. This is substantiated by the fact that they spent four years in school earning their degree, followed immediately by several years in the workforce before deciding they missed their calling. These candidates have learned the value of hard work and determination. Unfortunately, their career choice was not satisfying for them. Oftentimes they have learned that money is not the most important thing after all. They have discovered that although firefighters do not make a great deal of money (enough to be comfortable), a firefighter's job satisfaction rating is very high.

Once these candidates "round out" their education with fire science courses and a fire academy, a department quickly snaps them up. These candidates fit the profile perfectly of the older candidate who loves his or her job and excels in the fire service. Fire departments across the country have keyed into these candidates and hire them at their first opportunity. A candidate who has worked in another profession usually makes a strong firefighter, as he or she knows what it's like to work in a job in which there is minimal job satisfaction. Being a firefighter is a far cry from being trapped behind a desk in a cubicle.

A firefighter with experience as a drafter, computer technician or some other technical field brings a great new dimension to the fire service. Where once firefighters struggled with computers or prefire plans, the modern firefighter is able to create a computer generated mock up of a building. These plans include locations of hazardous materials, storage of company records as well as locations of fire department standpipe and sprinkler connections. The value that these drawings bring to an incident commander huddled around the command post is immeasurable. All of this because the fire chief elected to hire a firefighter with some computer experience.

Most firefighter candidates should aspire to complete at least an Associate's degree. Standard prerequisites such as math, English and writing are naturally required. Although it varies from college to college, the required courses usually include Introduction to Fire Science, Physics and Chemistry for Firefighters, Firefighter Safety, Fire Prevention, Building Construction, Fire Sprinkler Extinguishing Systems, Physical Fitness for Firefighters and Emergency Medical Technician (EMT).

Introduction to Fire Science teaches the student the basics of how the fire service works. It covers the difference between a fire engine and a fire truck, a captain and a chief. The course usually involves a class project in which the student is required to knock on the door of a fire station and research a firefighter's job description, regular duties and responsibilities throughout the course of his or her shift, the pay and benefit schedule.

Upon completion of the project, the student knows exactly what a firefighter does in the course of his or her shift and how he or she is compensated. This course is the basic framework that will give a student the confidence to walk into a fire station anywhere around the country and understand the basic terminology and operations that all fire departments follow.

Physics and Chemistry (sometimes called "fire chemistry") breaks down the chemical processes of how a fire starts, and most importantly, how it can be extinguished. The course covers the different classifications of fire and the basics of fire behavior. It covers the law of heat transfer and clearly delineates how a fire spreads throughout a structure or a forest. The more a firefighter understands the way a fire spreads, the better he or she will be able to combat and ultimately extinguish it. The course teaches the student to interpret and understand the labels present on all fire extinguishers.

Physics and Chemistry also provides a basic foundation for dealing with hazardous materials. Since there are so many toxic chemicals present in smoke (the byproduct of combustion), it is essential that a firefighter understands how it affects him or her. Firefighters are usually the initial response agency for hazardous materials incidents. This means that a firefighter must be trained to recognize the dangers and what needs to be done to minimize the adverse effects on the citizens of a community, their property and the environment.

Firefighter safety is a critical part of our profession. Statistics show that there is a strong probability that during the course of a career, a firefighter is going to miss time from his or her work due to a job-related injury. Being a firefighter is undisputedly one of the most hazardous occupations in the country.

> A candidate who has worked in another profession usually makes a strong firefighter, as he or she knows what it's like to work in a job in which there is minimal job satisfaction.

The Firefighter Safety course will teach students the importance of wearing safety equipment. It will examine firefighter death and injury investigations and seek to identify how each incident could have ended positively, instead of in tragedy.

Fire Prevention is also an important part of a firefighter's assignment. After all, most mission statements have a reference to preventing fires before they occur. A firefighter must be able to walk into a place of business and identify things that are in violation of the Uniform Fire Code. Our intent is not to write citations, but rather to get the business owner to rectify the potential fire causing violation. As a firefighter our salaries are paid by thriving businesses in the community. Our objective is to make the businesses "fire safe" so they can continue to employ the citizens of our community and contribute to the tax base. The Fire Prevention course will teach the aspiring firefighter the basics of the fire code as well as many of the most common violations encountered by firefighters. In addition, it teaches the student how and why firefighters have the authority to enter a business, make recommendations and ultimately mandate that a business comply with the established fire codes.

Building Construction is one of the most important classes a firefighter candidate will take. It is critical that a firefighter understand the basics of how buildings are put together, as many are killed or injured when buildings unexpectedly fall when subjected to fire. Students should be able to name all of the structural members used in the construction of a house or apartment building, as well as how large warehouses are constructed.

The more a firefighter understands what happens to different structural members during extreme situations, the more he or she can predict when a building will fail. With the advent of engineered trusses to support roof structures, many new buildings will predictably fail in as few as eight minutes when subjected to fire. This is usually about the time the firefighters have laid their lines and are making their initial attack on the fire. The Building Construction course will teach the student how to predict the longevity of each common building style.

Fire extinguishing systems are an integral part of how a firefighter attacks a fire. If a building has built-in systems that assist the fire department to extinguish the fire, minimize injury and reduce property damage, it is understood that the firefighters will be proficient in their use. The Fire Sprinkler Extinguishing System course teaches the student exactly how a fire sprinkler system operates and how to best use it to augment the firefighting efforts. Students will be tasked with designing a fire sprinkler extinguishing system for a large warehouse using industry standard formulas.

Emergency Medical Technician training is a critical part of a firefighter's training. Since the majority of a fire department's emergency responses are EMS-related (often as high as 90%), it is imperative that a firefighter is a proficient EMT. An EMT is able to splint fractures, take vital signs and place victims in cervical spine stabilization as well as perform a myriad of other responsibilities relating to patient care.

EMT is a 106-hour course. Although it may be offered within the context of the fire academy, EMT training should be taken before a student enters the fire academy. Since there is so much memorization required in the course, it is much easier if a student completes an EMT course prior to entering the academy. This is not an area an aspiring firefighter should be weak in.

A candidate who has a strong educational background will be much more attractive to fire departments. Although many firefighters are hired with little or no education, a candidate who has a degree will have a distinct competitive advantage.

Station Visits

Visiting fire stations is a critical part of the hiring process. You will get to know the details of the job, station life and that particular department's unique culture. Even more importantly, the firefighters will get to know you.

If you visit the stations early enough, before the department announces the recruitment exam, this can be a reality for you. If you show up with the testing crowd (all of the other candidates who show up once the department announces they are giving an entry-level exam), which often numbers in the thousands, your chances of getting to know the station crew and vice versa, are greatly diminished.

The actual testing may be done by the civil service or personnel department, with input from the fire department, but you can be sure the firefighters in the station houses have their fingers on the pulse of what's going on. As you can imagine, it is extremely difficult for firefighters on an interview panel to determine in a 20-minute interview if they want you as a member of the department for the next 30 years. Why not have firefighters who have gotten to know you, pitch you to the oral board?

It is important to understand that, good or bad, you are establishing your reputation the minute you walk in the door on your first visit. A positive opinion of you may make its way to the interview board. On the other hand, a poor first impression may also make it to the board.

It is hard to predict the best time to visit a fire station. In most agencies the phone numbers to the fire stations are not public record. It is up to you to go to the station and make contact. Since the firefighters are not sitting around waiting for you to knock on the door, there is a good chance you will miss them when you show up. This is especially true in high impact areas where the fire department experiences a high call volume. Ironically, the busier stations are the ones you want to visit. Since these stations run calls all day and night long, the younger firefighters are sent there to get experience. These are the ones who are the most current on the testing process since they went through it most recently.

Whenever you visit a fire station it is customary to dress nicely. Some candidates may elect to wear a suit and tie, while others prefer business professional (nice pants and shirt). Is a suit and tie overdressing for the occasion? Probably. How would you rather be remembered: as the candidate who showed up in a suit and tie, or one who showed up in a tank top and flip-flops? Believe me, the firefighters will remember you.

I know a candidate who showed up to a fire station wearing a suit and tie to inquire about the testing process. When asked if he was on his way home from work, he said that he was off today. When asked about his suit and tie, he told the firefighters that the job was very important to him and he wanted to make a good first impression. The firefighters were speechless. Imagine the help they gave him. (Remember, we want to hire people we like and who will cherish the job.) This candidate certainly made a strong first impression.

The same can be said for someone who comes by dressed in a less than professional manner. It is a poor choice to show up looking like you just came from the beach or the gym. It is important to remember that you are visiting our house. We are professional when we are called to your house. We ask that you show the same respect.

The firefighters appreciate (and expect) that you will bring an edible gift when you visit. An apple pie is my personal favorite. It keeps well as you are parked in front of the station waiting for the crew to return. Ice cream will melt and a banana cream pie (which is also a firehouse favorite) will get warm and spoil. A warm apple pie, however, appears to have just come out of the oven.

It is important to knock on the door or ring the bell, even if the apparatus doors are up. It is an invasion of privacy to walk in unannounced to a fire station and yell, "Is anyone here?" This would be similar to someone walking into your open garage and calling your name. It's just not good firehouse etiquette and is certainly not the positive first impression you were hoping for.

Once you have knocked on the door and a firefighter opens it, you should introduce yourself and ask if it is convenient for you to have a moment of their time. A pie in your hands will naturally increase the likelihood of them having some time for you. If it is not a good time for them, ask if you can make an appointment to come back later.

If you are turned down at the first fire station, go back to your car and open the map book and find the next closest station. In an urban area the next station may only be a mile or two away. In a rural setting you will have a further distance to travel. Once you have found the next station, repeat the process.

The best time to visit a fire station is between 3 and 5 pm. The firefighters have probably finished their inspections and are either working out or preparing dinner. The mornings are usually reserved for fire prevention activities or other fire department-related tasks. The late afternoon is usually less structured.

Once you have been invited in, it is important to explain that you are there because you are interested in getting hired by their department. Ask them if they know what to expect from the testing process. Generally speaking, the agency will give an exam on a fairly regular basis. Many departments have it in their city charter to have an "active" hiring list even if they don't have any projected openings.

It is important to have researched the department before going to the fire station. You can do this by visiting the fire department administrative office or by looking up the department's website. However you choose to do your homework, do not ask the firefighters how many stations the department has and how many calls they go on each year. This is a waste of THEIR time. You will quickly lose their interest and be politely escorted from the station at the first opportunity.

The best way to reinforce a positive first impression is to show that you have done your research. You can confirm your information by asking them if you can review it with them: "I understand that you have 23 fire stations and that your department runs 50,000 calls per year." The difference is that the firefighters see that you have taken the time to do your research and you don't expect them to do it for you. Again, it's a sign of respect. Their time is too valuable.

Some of the questions you should ask include but are not limited to the following:

1. How long is the probationary period?

2. What can I expect from the academy?

3. How is the relationship between the fire department and the community?

4. How many firefighters are going to be hired and how long is the eligibility list?

5. What desirable qualifications is the department looking for? (You will have already read the job description flyer, but you are looking for the "inside information.")

6. What are the strengths of the department?

7. What are the opportunities for advancement down the road?

8. What can I do to make a good impression on the oral board?

9. Is the fire department active in the community? (e.g. teaching first aid and CPR courses, public service day, CERT Training, etc.)

10. What do you like about the department?

11. What additional projects or assignments are firefighters able to get involved in? (e.g. fire prevention bureau, hazardous materials team, confined space or technical rescue and paramedic program)

12. What are the different areas of the community that the fire department services? (e.g. airport, marine, wildland interface, freeway, commercial, high-rise industrial, residential and beaches)

13. What special community projects is the fire chief planning to implement? (e.g. CERT Program, train a certain percentage of the community in first aid, CPR and AED, immunizations for the community)

14. What is important to the fire chief? (e.g. experience, education, mechanical aptitude, living in the community)

15. Is the city or county planning to add or eliminate fire stations?

16. For the new firefighters who have done well on probation, what qualities do they possess that have made them successful?

17. Where is the department headed in the future? (e.g. hazardous materials teams, weapons of mass destruction task force, immunizations for the community, add a BLS ambulance transport system)

18. What are some of the biggest morale boosters for the firefighters that have occurred in the last couple of years?

19. What projects has the department completed in the last few years? What projects are still in the works?

20. If you were in my position and you wanted to work for this department, what would be your next step?

21. Is there anyone else that you would recommend I speak to?

Most firefighters are very proud to be a member of their department. They want to be sure that the ones who follow feel the same way. A wise fire chief once told me, "The fire department ran well for 100 years before you became a member. You can bet it will run for another 100 years after you're gone. It's up to us to make sure we leave it in the hands of competent people."

As you can see, when you show up to the stations the firefighters are unofficially deciding if you are worthy of being a member of the department. It is imperative that you leave them with a positive first impression.

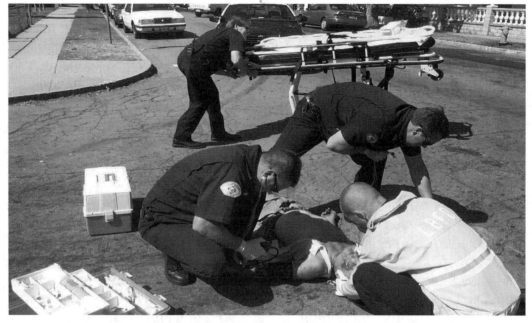

Ride-alongs allow a candidate to experience the fire service.

Ride-Alongs

Having the opportunity to ride along with a fire department engine company or paramedic unit is a monumental privilege. Due to liability reasons, most fire departments will usually deny civilian ride-alongs. Since the 9/11 terrorist attacks, many large metropolitan departments have banned civilian ride-alongs altogether.

The fire agency receives no benefit from allowing an individual to ride along. The potential for injury is extensive, in addition to the responsibility for another person while on scene. Being a firefighter is a dangerous occupation. Fortunately, firefighters develop a street sense in short order and are able to see when a situation is going poorly. Lastly, when traveling lights and siren, there is an increased risk of being involved in a traffic accident. Even though a fire apparatus is making a complete stop before proceeding against a red light, it is still a dangerous situation.

In addition to the added responsibility of having a ride-along in the field, it can be uncomfortable to have him or her in the station. The crew follows a routine and has a certain way of interacting with one another. This is all interrupted when a stranger is in the station. It is akin to having a houseguest in your home. You are always aware they are there and you cannot truly relax.

If a candidate is fortunate enough to procure a ride-along, he or she is in for a treat. Unless, of course, the crew experiences the "ride-along curse." It seems that invariably the crew will be extremely busy until the ride-along arrives. Predictably, once the ride-along is in the station, the calls for service immediately stop. When the ride-along leaves, the calls resume.

If you call a fire department and request a ride-along, the request will most likely be denied. An exception might be as a requirement for an EMT or a fire science course. In many cities, anyone who is competing for an entry-level position is prohibited from riding along, as it could give the candidate an unfair advantage in the process. If the candidate is discovered to have broken the rules, he or she is automatically disqualified from the hiring process. This is an easy way for the department to deny 90% of all requests.

The most likely way to secure a ride-along is to request it at the fire station level. If the captain of the station requests a ride-along through channels, the request is usually honored. A request from a friend, neighbor or relative would most likely be granted, but it's up to you to do the legwork.

Once your ride-along has been approved, it is important to find out what you should wear. Usually a nice pair of pants and a collared shirt would be sufficient, along with a clean pair of work boots, or tennis shoes as a last resort. When you arrive there is a good chance you will be provided with a fire department jacket.

Make sure you arrive at the fire station on time and always come bearing gifts. Traditional firehouse favorites include pie and ice cream. Mom's homemade apple pie or your favorite chocolate chip cookie recipe will also be appreciated.

Knock on the door when you arrive. Find out where you should park your car. In many of the areas you will be riding along, the neighborhoods will be a little on the rough side. You don't want to have your car stolen while you are on a call. Check in with the on-duty captain. He or she will probably have you sign a liability release form and will explain what is expected of you. It is important to find out what you are allowed to do. Most likely you will only be allowed to observe.

Find out what you are going to do on calls. Many departments state that you are to remain on the apparatus at all times. Many captains, however, will assign you to a mentor firefighter. It is your job to become his or her shadow, never leaving his or her side.

Find out where you are expected to ride on the apparatus. Figure out how the seat belt works and where it is located. Oftentimes they get tucked away, as the "extra" seat is usually not utilized. Place your jacket on the seat so you don't have to scramble around trying to remember where you left it. Under no circumstances do you want to have the crew wait for you.

Find out where the medical latex gloves are kept. Put a couple pair in your pocket and be sure to don them on every EMS response. Even if you are not allowed to do direct patient care, you may be handling the equipment as the firefighters are carrying it back to the rig. If it is a serious call, it is possible that they will need an extra pair of hands. You never know when you may get into the action. It is important that you are prepared.

Introduce yourself to each member of the crew; don't wait to be asked who you are. Get involved in the conversation if you are spoken to. If not, remember you are there to be a spectator, not the center of attention.

When a call comes in, get to your spot on the apparatus. Be the first one seat belted and be ready to respond. Stick with your mentor and don't leave his or her side. Do not engage in conversation with family members or bystanders. You are representing the fire department. It is important to act accordingly. If you are not sure of something, err on the side of caution. Remember that you are going to be allowed into situations that the general public is excluded from seeing.

During meals make sure you pay the going rate. If the firefighters do not accept your money, it is important to understand that they purchased additional food in anticipation of an extra person eating. Since we pay for our own meals, this extra money came out of the firefighters' pockets. It is important to insist that you pay your fair share, as you don't want to be a burden. If they refuse to take your money, don't force the issue, but be appreciative.

Be the last to sit down for meals and make sure to take the smallest portion. Follow the lead of your mentor firefighter. The minute he or she is finished and gets up from the table to do dishes, so too should you. If there is a rookie in the station, he or she will eat quickly and start the dishes immediately. In this case, you should be doing the dishes with the rookie.

If the crew plays a card game or rolls dice to see who gets "stuck" in the dishes, you should play if you are invited to do so. Even if you are lucky enough to win, you want to make sure you are still in the suds.

It is important to note that each member of the crew is evaluating you. They are listening to what you have to say, even if they don't appear to be listening to the conversation. You have already begun to establish your reputation. If the firefighters like you, they may invite you back. If you have established a rapport with the crew, they may be willing to draw up a plan of how you could become a firefighter. If you are really lucky, they may be willing to do a mock interview.

Being able to go on a ride-along is a privilege. It is important to remember that you are a guest in the firefighters' home. It is a burden to have you along. Do your best to make it easy for the firefighters to have you along and you will have a pleasant experience.

There are several firefighters currently employed by my department who showed up as ride-alongs several years back and decided after their experience in the fire station they wanted to be firefighters. They followed their dreams and are now the "mentor" firefighters for the ride-alongs. Their lives have come full circle.

Notes:
_____

_____

_____

_____

_____

Notes:

Be the change you want to see in the world.

– Gandhi

Role Model

As a firefighter, I recognize the need to be a role model in the community. Unfortunately, our society places great emphasis on being able to sing, dance, hit a baseball or shoot a basketball. If someone is famous, society tends to turn a blind eye to all of the baggage that often comes with these "superstars."

Granted, this is a book about getting hired in the fire service. There are so many parallels, however, between becoming a professional baseball player and a firefighter, that I felt the readers would benefit from the comparison. The challenges are comparable and there is no certainty that your hard work will be rewarded. These occupations both require that you pay your dues up front before you see any benefit. In addition, there are literally thousands of people competing for a few select positions. Each profession requires a person with a good work ethic, drive and determination, who is not afraid of competition.

First of all, it is important to note that I am not a baseball fan. I grew up riding my bicycle to Anaheim stadium to watch batting practice before the Angel games. After the second major league baseball players' strike, I am no longer interested in baseball. I do, however, appreciate a role model in the community, which is why I have so much respect for Shawn.

I had the great fortune of meeting Los Angeles Dodger center fielder All Star Shawn Green for breakfast. Shawn Green is the brother-in-law of a friend of mine who recently graduated from a junior college fire academy. After attending the graduation ceremony, I learned that during the height of the baseball season, Shawn took the time to attend the graduation. It was difficult enough for me to fit it into my busy schedule, but my schedule is nothing compared to a major league baseball player's.

Shawn attended the graduation and was simply introduced as Brian's brother-in-law, Shawn. I appreciate the fact that he took the time to go and never told anyone of his professional status. He believed this was Brian's special day. He did not want to take the focus from Brian's accomplishment.

On another occasion Shawn was hitting baseballs at the local high school while the kids were in class. The local firefighters were running around the field as part of their morning fitness routine. They were used to the high school kids hitting the ball and were amazed at the stranger crushing the ball. When they inquired about who he was, they were told it was Shawn Green of the Dodgers. They stopped running and stood behind the backstop to watch the show.

Shawn was in a groove and did not notice the audience. When a call came in over the portable radio, he realized he was being watched by the firefighters. He put his bat down and walked over to the firefighters and introduced himself as "Shawn."

The firefighters introduced themselves and asked if he was Shawn Green of the Dodgers. Shawn humbly acknowledged it. Instead of keeping the focus on himself, he asked about the Orange County Fire Authority. He mentioned that he had just attended his brother-in-law's graduation from the fire academy. Knowing that Orange County Fire Authority was going to be testing soon, he asked what Brian could do to get a job. Of course, the firefighters gave him their number and said Brian could call them for help.

After hearing these examples of how Shawn conducts himself, I asked Brian to set up a breakfast meeting. I was pleased to learn that he was willing to meet with me. The following is a dialogue between us. The important points to remember are the many similarities between a major league baseball player and a firefighter.

Shawn Green is 30 years old and has been in the "majors" for 10 years. He is a two-time All Star and a Gold Glove winner. He played high school baseball at Tustin High School, where he earned straight A's. He signed a letter of intent to play baseball at Stanford. He always planned to be a professional baseball player.

What do you like best about your job?

Similar to being a firefighter, I enjoy the camaraderie in the clubhouse. We spend a great deal of time together. We play 162 games in 181 days. Chemistry in the clubhouse is immensely important. I have never been in the playoffs.

*It is amazing what you can accomplish if you
do not care who gets the credit.*
— Harry S. Truman

Each year there are one or two teams that have great chemistry. Last year it was the Angels and they went all the way. They did not have all the talent, but they had the chemistry.

To what do you attribute your success?

Hard work. I have to work hard. Things do not come easy for me. I remember as a rookie standing behind the batting cage and watching guys crush the ball. I thought to myself, "What am I doing here?" These guys are big and strong and they can crush the ball. They are physical specimens and I am not as physically gifted. During game situations, however, they did not produce. One by one they disappeared. Some did not have a strong work ethic, while others did not perform when it counted.

I have to keep on top of my mechanics when it comes to hitting. It's a constant battle. I lift weights and spend a lot of time in the batting cage.

Do you see yourself as a role model?

Definitely, all athletes are. It's important that all kids have role models. It's too easy to be selfish when you make a lot of money and have a lot of fame. Some people just go crazy. This is very comparable to being a firefighter. There are times when you are in a hurry and someone wants to say hello and shake your hand. If you don't take the time, he or she will remember it forever. Sometimes it can be difficult. It is a little easier for me since I live in Orange County and play in Los Angeles. Most people don't recognize me.

What do you have to say to people who aspire to be in your position?

I know what Brian, my brother-in-law, is going through to become a firefighter. There are many similarities to my situation. The best advice I can give to anyone who is chasing a dream is that it is important to love what you are doing. If it feels like work, it may not be the right thing for you. You should enjoy 2/3 of what you are doing. It's not about the money and fame.

You must seize the opportunity to do everything to stack the deck in your favor. You have to be fortunate, but nothing comes without hard work. You also have to have the right doors open for you.

What was your biggest setback?

I recall playing in the minor leagues. I got called up to the majors and did not produce. I stayed up for a month and was sent back down. It was really tough to get called up and have everyone evaluating you and then fail. You worry about being a 30-year-old rookie with a family and never being able to stay in the major leagues.

I also remember playing for the Toronto Blue Jays the year after they won the World Series. The Manager did not like having young players on the roster. The following three years we did poorly. He blamed the demise of the team on me. He told me I couldn't hit left-handed pitchers and I was not a good outfielder. In short, I became his whipping boy for three years. I kept my head down and continued to work hard. I showed up to work early and left late.

Ultimately, he was fired and I stayed on. The next year I won a Gold Glove and was voted to the All Star team. The important thing is that I didn't get discouraged and give up. I kept my head down and worked hard.

How do you deal with self-doubt and what advice do you have for others on how to deal with it?

Everyone experiences self-doubt, no matter what position in life they hold or aspire to. Every time I go to spring training I have performance anxiety. Everyone experiences it, from the best player to the worst one. Everyone has slumps. When you are playing well you worry about going into a slump. When you are playing poorly, you worry about getting out of your slump. Ironically, I find it easier as I get older. It's easier to get out of a slump since you have been able to climb out before. It is important to have the confidence to know that you can overcome adversity. People don't reflect on changes they need to make along the way when things are going well. They only evaluate things when they are going poorly.

If it weren't for baseball what would you do?

I signed out of high school to play at Stanford. I knew I would be a major league baseball player. In the off-season I took classes toward my degree. If it weren't for baseball I would have been an Orthopedic Surgeon. I believe people need to figure out what they want to do and chase their dream. It's not my personality to go through life without direction. I believe in setting goals and working to accomplish them.

How do you define success?

Success can be different for everyone, depending on his or her challenges. I believe it's important to accomplish what you set out to do. It is imperative that you learn to accept the struggles and failures along the way.

Shawn Green is an example of a very public role model whose determination, focus and hard work have enabled him to achieve his professional goals. His willingness to learn from those who are more experienced and work as a team player have helped him reach his goals and undoubtedly make him well-liked by his teammates. His work ethic, humility and appreciation for his job make him an ideal role model for others to follow.

Notes:

How to Succeed in Fire Technology Courses and the Basic Fire Academy

Steve Prziborowski is a Battalion Chief with the Santa Clara County (Los Gatos, California) Fire Department and the Fire Technology Director at Chabot College in Hayward, California. Steve has agreed to share his experiences and thoughts with future firefighters.

Becoming a firefighter is not an easy task. It takes a great deal of perseverance, patience, persistence, dedication and good old-fashioned hard work. Nothing in life comes easily, especially when you want to have one of the best careers a person could ever wish for.

How long will it take to become a firefighter? That question cannot easily be answered because it really comes down to "what you give is what you get." Not every person who starts out intending to become a firefighter ends up becoming one. For that matter, not every person who goes to medical school becomes a doctor; nor does every person who goes to law school become a lawyer. See where I am going with this?

I cannot guarantee that you will ever reach your dream of becoming a firefighter. What I can guarantee is that if you never give up on obtaining your dream, your odds of succeeding increase greatly. On average, I would say it takes anywhere from three to seven years to become a full-time paid firefighter. Some do it in less time, some do it in more time and some never get the chance to do it at all. I have known people to take 10 to 15 years to become a firefighter. I have known people to give up after only a year.

What's the moral of the story? If you give up, you give up your dream. If you continue pursuing your dream and continue doing whatever it takes to achieve that dream, continuously working on improving your weaknesses, keeping up your strengths and preparing yourself to be the best candidate that you can be, then you stand the chance of actually achieving that dream!

Once you have determined that your dream is to become a firefighter, then it is time to put your money where your mouth is and start preparing yourself on a full-time basis. It is a full-time job just getting the job!

Beginning the Process

At the point you determine you want to become serious about becoming a firefighter, you should enroll at a junior college that offers fire-related education and begin taking classes to work towards your two-year degree in Fire Technology (a.k.a. Fire Science).

I would strongly suggest that you plan to complete your two-year degree in Fire Technology (as opposed to just going there for your EMT certification and Firefighter 1 Academy certification). While it is true that most fire departments do not require a two-year degree (or higher level of education) to take their firefighter examination, you must realize that a good number of your competitors will already have at least a two-year degree or be close to finishing it. When an oral board has seen a week's worth of candidates who have at least a two-year degree, how would you like to be one of the people they interview who does not have a degree, does not appear to have any desire to complete one, or even further your education?

Formal education in the fire service is here to stay and the demands and requirements are only going to go up from here, even for entry-level firefighter positions.

Successfully Completing Your Fire Technology Degree

In the ten or so years I have been instructing Fire Technology and EMS-related classes and the two and a half years I have been coordinating the Fire Technology program at Chabot College, I have had the opportunity to interact and work with countless students aspiring to become firefighters. The fire technology program at a college is usually one of the more popular programs;

if you have ever taken an entry-level firefighter exam, you have seen that the number of people applying exceeds the number of open positions.

Do not let the supply and demand factor discourage you in your pursuit of your dream. I say that because among all of the students that I come in contact with who say they want to become a firefighter, many of them are neither prepared nor motivated to jump through all of the hoops required. Experience has shown me that out of every class of 50 students (the typical number of students in the beginning fire technology classes at the college), there are only about 10 to 15 who will probably become firefighters. That is my rough estimate based on my perceptions of performance in areas such as grades, attendance, level of motivation, level of dedication, etc. Remember the statement that "perception is reality."

I remember when I started out taking fire technology classes at the college. My best friend and I had discussed this same subject. We agreed that out of the 50 or so fellow students in the class who were starting out the program with us, we did not appear to have that much serious competition. While there were a large number of people we initially perceived to be competitors, we quickly learned that most of them were not as motivated or dedicated to becoming firefighters as we were. While we were slightly older than many of the students (we were 24 years old), we knew that we had an excellent chance at becoming firefighters if we stuck to a basic plan and did not give up.

Our basic plan was to complete our two-year degree in the quickest time possible while still having a social life, working (can't forget about having to pay the bills), taking firefighter examinations, making the time to do some volunteer work and actually putting some time and effort into each of those firefighter examinations so that we could do our best and rank as high on the list as we possibly could. It is possible to complete a two-year degree in 1 ½ years if you take as many classes as you can each semester, attend school year-round, remain focused and have some luck as well (e.g. not finding all of the classes full, not finding classes cancelled, etc.).

How do you become one of those top 10 to 15 students in each class who has the greatest chance of actually becoming a firefighter (and finishing your two-year degree)? The following are some suggestions that will help you get through a college fire technology two-year degree and recruit academy, through your probationary period and even through your career as a firefighter.

1. **Time Management/Attendance:** Always show up to class on time and return from breaks on time. Nothing frustrates an instructor more than having students walk in during a lecture, because then they have to repeat themselves (because you know those students will ask a question that you have already gone over while they were not present). Being late also disrupts the flow of the class because the instructor and the students get distracted and lose their focus. Do you think a fire department is going to let you into their written examination after the starting time? Doubt it; there are hundreds if not thousands of other candidates who made the effort to be there on time and obviously want the job more than you do (remember, perception is reality).

   Many students have trouble attending every class session. If you are fortunate enough to get hired by a fire department, they will expect you to show up every day ready to go to work and do the job you're being paid to do. Start doing what it takes to keep yourself healthy and in top shape so that you can attend every class session. Missing any class session can lead to termination from any of the classes because of mandated time requirements from the State and/or the College.

   Excessive tardiness can also lead to your termination from any of the classes and will not be tolerated.

2. **Common Courtesy:** This appears to be a lost art. If you are going to be late for class, make every attempt to contact the instructor to advise him or her of your situation. Instructors (as well as supervisors) really appreciate it when they are "kept in the loop" and are not blind-sided by things. If I know a student is going to be late because he or she contacted me in advance, I can be flexible. However, when you show up late to class (unannounced) when a test or an assignment is due, don't expect to be able to do a make-up test or obtain full credit for the assignment.

3. **Maturity:** Realize that your instructors may be on your oral board panel in the future, or you may be working for them at some point in your

career. If nothing else, you will hopefully be using them as resources to assist you in your pursuit of becoming a firefighter. If you are acting immaturely (as many students do), what type of message does that send to your instructor and the other students (your competition)?

4. **Respect:** Show your fellow students and instructors the respect they deserve. There are always a few students in every class who seem to forget this concept. They talk while instructors are speaking, make fun of other students and argue with the instructors. The list goes on.

5. **Cellular phones:** Leave them at home or in your car. Nothing disrupts a class more than a ringing cell phone (except for the person actually answering it and talking to the caller). Do you think you are allowed to pick up a cell phone while working a medical call as a firefighter? Why is that different from being in class?

6. **Turning in assignments on time:** I've never heard an instructor complain about a student turning in an assignment early. Most of them would probably appreciate it because it would allow them to grade it early and get it out of the way.

7. **Written test taking skills:** Most classes, academies and fire departments expect you to keep an 80% average on your written tests. Every semester we terminate students from the academy, the EMT program and other fire technology classes because of their inability to keep an 80% average on their written tests. I get sick of hearing "I don't do well on written tests. Let me do the hands-on skills and I'll be perfect." Firefighters need to be perfect at the hands-on skills and near perfect on the written tests. Do what you have to do to increase your written test scores. Increase your study time. Study the right material, study with a group and do whatever it takes. Like it or not, a firefighter will be taking written tests for the rest of his or her career, whether it is for a promotional exam, continuing education, or annual mandated training.

8. **Attitude:** Keeping a positive attitude is a necessity to successfully become a firefighter (and remain employed as a firefighter). One of my students recently told me that he has a problem with authority figures. I politely advised him that if that is true, then he should probably look for

another career. He either thought he could get by, or that the fire service would allow him to be that way. Regardless, we are still a paramilitary organization for a reason. Many students have poor attitudes for a variety of reasons. The bottom line is that it is easier to give them the benefit of the doubt if they have a positive attitude.

9. **Ability to receive constructive criticism:** I'm not sure if it's just the younger generation of today or my getting older, but it seems that many students do not like to hear constructive criticism. If you cannot take and process constructive criticism, how can you improve your performance? Most instructors and fire service professionals are sick of hearing excuses from students and probationary firefighters about why something happened (or didn't happen). Remember that nobody is perfect, even you!

Find out as many of your weaknesses as possible so that you can improve on those weaknesses and be successful! Accept graciously everyone's opinion when they are discussing your performance. While nothing says you have to agree with what they are saying, do give them the respect they deserve for the position they are in and actively listen to what they are telling you. You might learn something about yourself that you didn't know! You might also learn something that you realize is a pattern. If multiple people are telling you about a weakness, then take the time to consider that maybe you are the one who is wrong, not them!

10. **Being responsible/accountable for your actions:** One of the things that bugs me is students making excuses. Own up to your actions, take responsibility for what you have done (or have not done) and honestly say, "I screwed up." We've heard almost every excuse in the book, so don't sound like a broken record. I say almost every excuse because every semester we hear a new one. Personally, I don't care about all of the traffic that was on the road this morning. Don't you think I was probably aware of it because I also had to drive to the college? Regardless, you should have left yourself ample time to get to school, find parking, find the classroom and handle any issues that came up. When your relief at the fire station is expecting to go home at 8:00 a.m. and you are late, how do you think he or she is going to feel? I don't

appreciate it when I have to hold over longer than I have to every day (unless it is pre-arranged or truly an emergency – which most of the time it is not). I've got places to go and things to do. Plan to get to class or the fire station at least 15 minutes early (many firefighters get to the station 1 to 1 ½ hours early, just to beat the traffic and ensure they have plenty of time to spare).

If you forgot to do something or you did not properly prepare for something, just be accountable for your actions. Tell me that you screwed up and you won't let it happen again, or ask if there is anything you can do at this point to make the situation better. Make sure you are sincere and really mean it. Do you think the person you are relieving at the fire station (who has to immediately go to the airport to catch a plane for vacation) really cares that there was traffic or that you forgot about coming in on time (or early)? NO!

11. **Following directions:** Last, but definitely not least – if you cannot follow simple directions, how are you going to do so while on the fireground in the middle of a life-threatening emergency? The answer is you're not! Some students think I'm too tough on them or that I'm anal retentive (I like to think of it as "detail oriented") because I am so demanding and I have high expectations of them, or because I expect them to completely follow the directions I give them. If I didn't want them to succeed as firefighters, I wouldn't be such a stickler on this subject.

Many firefighters today are injured and/or killed because of their inability or their lack of desire to follow directions. This is unacceptable! Let's stop this vicious cycle! There is a reason the fire service is still "paramilitary." Around the fire station, there really isn't that much of a need to be militaristic in nature. However, on the fireground, there is a definite need for a militaristic approach to operations because lives are potentially at risk (ours and the citizens). Not following directions on the fireground can lead to the serious injury or death of a citizen or firefighter.

Successful students follow directions well. It amazes me how many students are not able to follow the most simple directions. In every class I try to impart the importance of following directions. On the first day of every semester, I pass out contracts for the students to sign indicating that they agree to read and follow the directions contained within the syllabus. Failure to follow directions can lead to their termination of class time for the semester. Every semester I inform multiple students that they have been dropped from a class because of not following the directions they previously agreed to.

Successfully Completing the Recruit Academy

At Chabot College we offer a firefighter 1 academy, similar to many other fire technology programs across the United States. Our firefighter 1 academy starts out with 32 students every semester. We run two academies per year: one in the spring and one in the fall. Of the 32 students who begin the academy, it is not uncommon to only graduate 15 to 25 students. Many of the students who have not been allowed to complete the academy for various reasons have successfully completed it after coming back a second and/or third semester. Why are we losing so many students and what can you do to successfully complete a firefighter 1 academy the first time you take it?

In addition to the above-mentioned suggestions to help you complete your fire technology degree, here are some specific suggestions to help you successfully complete the firefighter 1 academy.

1. **Manipulative skill ability:** A big reason students get dropped from the firefighter 1 academy is the inability to successfully complete the various firefighting skills they are continuously tested on. It seems that the tools and equipment that provide the most challenges for students are the wooden ladders, ropes and knot tying. Many times they can do the skills without a problem, but they cannot do them without performing any safety violations or within the expected time frame. Many times the reason for the inability to complete the skills within the expected time frame is the lack of physical conditioning or preparation.

2. **Physical conditioning:** Many candidates are not prepared to perform to their fullest capabilities because of their lack of physical conditioning.

Enrolling in fire science courses and completing the basic fire academy will greatly enhance your chances of realizing your dream of becoming a firefighter.

Many firefighter 1 academies expect the recruits to do an hour of physical fitness every day. This is time dedicated just to running, weight lifting and aerobic conditioning.

Additional Tips to Successfully Become a Firefighter

1. **Take firefighter tests:** Start taking as many firefighter entrance examinations as you qualify for. Every city that has a fire department usually has its own testing process that occurs once every two to four years. For example, if you want to work for the Oakland Fire Department, then you will have to participate in their firefighter examination process.

   How do I find out which fire departments are accepting applications and what are the requirements to become a firefighter with that department?

Subscribe to firefighter examination notification services. They are worth every penny of their price. They save you the time and effort of calling each fire department and asking when they are next hiring. These services provide websites with notifications of when departments nationwide are accepting applications.

Contact individual fire departments and their respective city (or county) personnel (or human resources) offices. To find out how to contact them, you can either do a search on the Internet or look in the blue pages of the phone book (which are the government pages). Typically the personnel department (or human resources department) for a jurisdiction handles the testing process for positions within the fire department. Ask them when they will next be testing for the position of firefighter, what the qualifications are to become a firefighter, whether they accept interest cards (if so, can you leave your name with them so you can be notified of their next exam?) and any other relevant questions you may come up with.

2. **Understand the requirements to apply for firefighter positions (and make sure you meet them or have a plan to meet them):** Some of the requirements to be able to file an application at various fire departments can include:

    - Minimum age: 18 or 21 years
    - Valid State Driver's license
    - Excellent physical fitness
    - Not currently a smoker
    - Bilingual ability (in any language)
    - Clean driving record
    - Current EMT certificate
    - Current Paramedic license
    - Firefighter 1 academy completion certificate
    - Firefighter 1 State certification
    - Two-year degree in fire technology (or closely related field)
    - Certificate of Achievement in fire technology

- 15 units of fire technology related classes from a junior college
- One year experience as a paramedic
- One year experience as a volunteer firefighter
- One year experience as a paid firefighter
- Current CPR certification
- Instructor Certifications in:
    - o CPR
    - o First Aid
    - o EMT
- Specialized EMS certificates such as:
    - o Advanced Cardiac Life Support (ACLS)
    - o Pre-Hospital Trauma Life Support (PHTLS)
    - o Basic Trauma Life Support (BTLS)
    - o Pediatric Advanced Life Support (PALS)
- Specialized training certificates from the State Fire Marshall's Office in:
    - o Hazardous Materials Decontamination, Awareness, Operational, Technician and Specialist
    - o Technical Rescue subjects such as low angle rescue, swift water rescue, confined space rescue, rescue systems, etc.

If you have been testing for firefighter positions for any amount of time, you have probably realized that fire department minimum qualifications vary greatly. Even though many of the above items are not normally listed as minimum qualifications to take the examination, I have seen all of them at one point or another as either a minimum, desirable, or highly desirable qualification. Your goal should be to try and get as many of them under your belt as you can before you take the examination, or be very well qualified if a department is listing those items as desirable or highly desirable.

Some departments require one or more of the above qualifications. It is feasible that you may qualify to take a firefighter entrance exam even before you start taking classes at a junior college. I was eligible for the City of Hayward's Firefighter examination even before I started taking fire technology classes at Chabot College! I definitely filed my application just so I could see what the process consisted of (and also had a slim hope of maybe getting a job offer). I was one of about 3,000 other applicants who were allowed to take the test because, if I remember correctly, they only required applicants to be 18 years old and have a high school diploma or equivalent. Probably the best thing to come out of the testing process was that I failed the first step, the written examination. I failed it! Looking back, it was an excellent eye-opening event that humbled me and made me realize I didn't know as much about being a firefighter as I thought I did. After that test, I properly prepared myself so I never failed another written test. I actually prepared so well that I was getting scores in the high 90's on my written tests.

Is it realistic to get hired as a firefighter without any training or education? While it is not realistic, it is not impossible. However, remember that having the above qualifications only allows you to participate in the hiring process; they do not guarantee your success.

The bottom line is that the more tests you qualify and apply for, the better chances you have of getting a job as a firefighter!

3. **Understand as much as you can about the job of a firefighter:** Start educating yourself on the job of a firefighter and the operations of a fire department so that when you are talking to firefighters, visiting fire stations and participating in the various events of the hiring process (such as the oral interview), you can speak in an educated way and actually sound like you know what you're talking about. Knowing the difference between an Engine and a Truck is important PRIOR to the oral interview. One way to learn as much as you can about the job of a firefighter and how the operations of fire departments can be similar, yet different, is to start visiting the websites of fire departments.

Summary

Remember that nothing worth having in life is going to come easily. It is up to you to remain positive, focused and motivated to continue doing what it takes to become a firefighter. There are going to be many frustrating and disappointing moments while testing; the key points are that you recognize your weaknesses, are open to constructive criticism and continue to pursue that dream of becoming a firefighter. Once you give up, you let someone else take your spot riding on that fire engine that you dreamed of riding on!

Biography for Steve Prziborowski

Steve Prziborowski is a Battalion Chief with the Santa Clara County (Los Gatos, California) Fire Department and has been in the fire service for 15 years. He is also the Fire Technology Coordinator at Chabot College in Hayward, CA, where he has been instructing fire technology and EMT courses for over 10 years. He is a state certified Chief Officer, Fire Officer, Master Instructor, Hazardous Materials Technician and state licensed Paramedic. He has an Associate's degree in Fire Technology, a Bachelor's degree in Criminal Justice and a Master's degree in Emergency Services Administration.

He also publishes a free monthly newsletter geared toward better preparing the future firefighter for a career in the fire service, "*The Chabot College Fire & EMS News*," that is available on his website at www.chabotfire.com.

Can people count on you?

Friends and Acquaintances

Fire departments have the luxury of choosing any candidate they feel would best represent their organization. To an outsider, if your friends are of questionable character, then so are you. You have heard the old saying, "Birds of a feather flock together." If your friends are getting into trouble, then, potentially, so are you. Even if you are not a party to illegal, immoral or unethical acts, you are going to be associated with these activities. Keep in mind that even if it is not illegal, it may not be the right thing to do. Why would a fire department choose to take a chance on someone with questionable character, when there are so many candidates to choose from?

Why is your selection of friends so important to the fire service? If your friends have questionable values and exercise poor judgment, it is probably a matter of time before they get into trouble with the law. Legally, you may be charged as an accomplice to a crime if you are just present, not even participating. Furthermore, you may be charged with possession of drugs if you are riding in a car with a friend who has drugs. The media would not be interested in your friends, but it would make the headlines to hear that a firefighter was arrested. Imagine the reflection on the fire service as a whole if a firefighter were arrested for burglary or on a drug-related charge.

The black eye to the fire service in general is tremendous, but the liability to the agency is even greater. Now the fire department has to deal with the public wondering if all of the firefighters are committing crimes. Obviously, this is not the image the fire service is looking to project. We go to such extremes to enhance our image in the community that we would never consider hiring anyone who could potentially bring discredit to the fire service.

Reflect back to a time when a professional athlete was caught doing something illegal. Unfortunately, you don't have to look too far. The damage control required to restore the team's reputation is astronomical. These "role models" have ruined the sport for all of the ones who do a good job. The fire department simply will not tolerate any firefighter bringing discredit to the fire service. It is much easier to keep a good reputation than to restore one.

Firefighters have a certain dress code and appearance. In short, the rules and regulations are pretty clear-cut. They require a conservative hairstyle and overall clean-cut appearance. If your friends don't fit the model, you may want to look for another group of friends. I am not saying that if your friends do not wear "nice" clothes each day they are not good people. I am saying that there is a standard. The appearance of your friends can be an indication of your own appearance when off-duty. If you do not want to conform to the paramilitary style of the fire service, it may affect your chances of getting hired.

When I decided that I wanted to become a firefighter my junior year of high school, I was forced to "abandon" my circle of friends. They were not bad kids; they were just average "curious" adolescents. I believed they (we) were involved in questionable activities that would reflect poorly on me in the future. Most importantly, I found that when I was around them I was both an instigator and follower. In short, I became someone I believed was not a good representative of the fire service.

In order to achieve my personal goals, I had to make some hard decisions. It was very difficult, but I knew I had to remove myself from the group. Peer pressure is never easy, especially at that age. I found it best to be honest with them and explain my reasons.

In time, I made new friends. To this day I still communicate with my old high school friends, but I don't have much in common with them as our lives have gone in different directions. They are all successful in their own careers and personal lives, but we have grown apart.

There is a chance that I could have hidden my relationships with less reputable friends, but to me it was not worth the risk. When or how does the fire department come up with the information about your friends? The background check is designed to do just that. A background investigator employed by the fire department will literally knock on your neighbor's door with a Polaroid picture of you and start asking questions. If the report from the people who see you on a regular basis is less than glowing, there is a good chance you will be passed over.

Very rarely do you hear that "Bob is a good person, he just chooses poor friends." Well, that's not entirely true. I would expect to hear that from Bob's mother. The real question is whether Bob is the follower or the leader. Either way, his selection of friends and therefore his judgment, is suspect. If he cannot make good choices about his friends, why would we expect him to make good decisions while wearing our department's badge and uniform? Remember, we have the luxury of hiring whichever candidate we choose. Why would we choose someone who exhibits poor judgment?

Friends are an important reflection of who you really are. Yes, you are responsible for your friends' actions while you are together. Take a close look at your friends and make sure you want to stake your future on them, because you really are.

Your friends should be a positive influence on you. They should encourage you to reach your goals while building your self-esteem. This of course is a two-way street. You should be able to share your dreams and goals with them. They should do anything to help you reach them.

A friend is someone you can call at any time of the day or night and ask for help. A true friend is the one who will drop anything to help you, even when it is inconvenient for him or her. Everyone is a friend when the beer is flowing and the money is rolling in. A true friend is someone who sticks by your side and lifts you up when you are down.

A true friend will tell you something that's tough to hear. It is easy to tell friends things that will make them feel good (and you should), but it is also important to tell them things that may hurt. A true friend will tell you the truth no matter how difficult it is. One of the most important parts of having good friends is to be a good friend. You must go out of your way to help friends out when they are in need. To receive you must first give. When a background investigator interviews these friends, they can say what an outstanding person you are and how you are always ready to lend a hand.

If you live your life with code and honor,
the gods will repay you.
– Deputy Chief Alan Patalano

Private Interview Coaching

One of the most overlooked aspects of becoming a firefighter is the oral interview process. While not all fire departments incorporate a structured interview into the hiring process, the vast majority do. It may be found as an initial part of the exam process, or the fire chief may include a one-on-one interview as the final step in the hiring process.

The interview is usually weighted the highest percentage of the applicant's final score. It is not uncommon to have the written and physical ability tests weighted "pass or fail," while the interview accounts for 100% of the final score.

Most fire departments use the same basic questions for the interview process. While there are certainly variations from agency to agency, the underlying theme of the questions is the same. This is easily explainable by the fact that all fire departments are looking for the same basic candidate. This "ideal" candidate has strong ethical and moral values, is physically fit, gets along well with others, is a team player, has mechanical ability, communicates well and is able to learn and retain written and practical information. There are numerous other dimensions that are beneficial to the fire service, but these are the general traits desired.

Oftentimes the questions will attempt to put candidates in a "no win" situation. In fact, the candidates will be informed that there is no right or wrong answer; rather, the board is simply looking to evaluate their thought process. While the candidates may be told this, don't fool yourself into believing this. There are usually automatic failure points that, if not addressed, will result in a low score, or worse yet, a failing grade. If a candidate ventures into "unfriendly territory," he or she may find himself or herself eliminated from the process completely.

If the board is simply trying to get to know a candidate and not testing his or her knowledge of the fire service, how can he or she improve an interview score? The answer is simple: private coaching.

It is similar to learning to hit a baseball. Can a person learn independently how to swing a bat and make contact with a baseball hurled toward home plate at 75 miles per hour? The answer is probably yes. Could the learning curve be drastically shortened by having a coach teach you the fundamentals? Of course! Imagine someone with experience showing you how to hold the bat, where to stand in the batter's box, how to anticipate the location of the pitch, and most importantly, how to make contact with the pitch. Again, it is possible to learn these fundamentals on your own, but a competent coach can shorten the learning curve significantly.

The same thing can be said for an interview coach. A coach who understands the interview process, and who has sat on the other side of the table as an evaluator, can provide insightful information. His or her years of experience, coupled with the knowledge of how the interviews are graded, will be invaluable to someone who is trying to learn the process.

Is it possible to get hired without the aid of a coach? Of course! In fact, the vast majority of firefighters across the country are able to get a job without the aid of an "expert." Some candidates have the gift of gab or the unusual knack of knowing how to take a fire department interview. For these candidates, private coaching may not be necessary. For candidates who are struggling with the interview process, private coaching may be just what is needed to make the difference between getting hired and having to wait several years until their dream department tests again. Consider how the testing process actually works and how closely the candidates score. There are usually no more than a few points between the candidate who gets hired and the ones who were oh-so-close.

Ensure that the person who is giving advice has the credentials to back it up. Although someone may declare himself or herself to be an "expert," he or she may not be. It is important to note that there are people who conduct personal interview classes and sell information to aspiring firefighters who, in my opinion, are giving bad advice. It is a tough situation for a candidate, because he or she does not know that the advice being given is poor.

Any information is helpful, but you as the candidate have to "buy in" to what your coach is telling you. If it doesn't make sense, the advice may not be sound. If you are uncertain of some of the advice you were given, bring it to the attention of your fire science instructors, the local firefighters, or someone you respect. As always, it is up to you to determine if you agree with the information being provided.

Hiring Process

Written Examinations

Written examinations are usually used as qualifiers to thin out the number of applicants. They are best associated with a form of general knowledge tests where a candidate's basic knowledge is tested. They are usually graded as a pass or fail and generally comprise a relatively small percentage (if any) of the candidate's final score. Oftentimes the minimum passing score is well above the traditional 70%. The reason the agency will not reveal the minimum passing score before the candidates take the exam is because they want to make sure they eliminate enough candidates to make the numbers more manageable.

If a department has 500 applicants for five positions, they can safely figure on a 10% no show for the exam. Out of 450 candidates, the department would like to have 200 move on to the physical ability exam.

If the department has a minimum score of 70% to pass the written examination, they may end up with 350 candidates. By adjusting the passing rate to a 75% they end up with 300 candidates, while an 82% gets their target number of 200 candidates.

Written examinations seek to establish a minimum level of math, English, reading comprehension and mechanical aptitude. Some examinations will also test a candidate's ability to read a map and follow oral instructions.

Since most written examinations are a general competency test, it is possible for a candidate to identify his or her limitations and improve upon them. The best way to prepare for a written examination is to eliminate known areas of weakness. For example, a candidate who is weak in math can enroll in a basic math course, while a person who is challenged by English can enroll in an English course. For those who do not have the time to enroll in a formal class, there are numerous books to help them prepare. While some of these books are geared specifically for fire department exams, a high school SAT preparation book should also help.

Another area fire departments are testing in the written examination is the psychological profile of a candidate. They try to determine how a candidate will interact with other firefighters for long periods of time. The difficulty with these exams from a candidate's perspective is that it is difficult to pinpoint the right answer and they are impossible to study for. Since there is no clear-cut right answer, there is no clear-cut algorithm to improve test scores.

Mechanical aptitude is another area that is difficult to learn from a book. Although there are some tricks of the trade that can assist you in learning how gears and pulleys work, the best teacher is always hands-on experience. A high school auto shop class will certainly help a candidate learn to work with his or her hands.

Since an agency will generally not give out the results of the written examination, it is difficult for a candidate to learn how he or she can improve a score. The best way for a candidate to know where he or she is losing points is to reflect back to the exam. When you turned the page and let out a groan because of what you saw on the page, there is a good chance that this is one of your areas of weakness. Whatever the area, you need to improve upon it. If it is challenging for you, it is probably difficult for the person sitting next to you. Since the passing scores are very tight, it is imperative that you don't give up any ground on any phase of the examination.

One way to prepare for a fire department written exam is to read and study my book, *Smoke Your Firefighter Written Exam*. While there are other self-help books devoted to helping aspiring firefighters, they only provide questions and answers. *Smoke Your Firefighter Written Exam* goes beyond providing sample questions and answers. It teaches the reader the basic rules and principles behind the question; in other words, how to solve each complex problem. Each section begins with an overview of how to solve the problems. The reasoning behind the correct answer is presented in clear, easy-to-understand language.

If the department requires a fire academy graduation certificate in order to take the examination, there is a good chance the written exam will be based on the International Fire Service Training Association (IFSTA) curriculum. These exams are usually taken directly out of the IFSTA Essentials textbook. Frequently candidates are not briefed on the contents of the examination until they open the first page of the test book. The reason for this is clear. Fire departments are looking to hire candidates who have kept up on the knowledge they have learned in the academy, rather than ones who can cram for an exam. The best way to prepare for these exams is to regularly review your academy notes, and more importantly, read through the IFSTA Essentials. I recommend reading a chapter a week to keep the material fresh in your mind. After all, since this is your chosen profession, it is incumbent on you to learn and retain the information that is part of your job.

Another way departments are looking to reduce the number of candidates moving through the process is via a series of simulated situations that a firefighter might encounter during the course of his or her duties. Instead of a formal entry-level interview, these scenarios take the place of the written examination. A sample of these scenarios might involve an interpersonal conflict with another firefighter in the station, or dealing with an unhappy citizen on an EMS run.

The best way to prepare for these examinations is to understand how to handle situational questions. I have outlined over one hundred questions in my book, *Smoke Your Firefighter Interview*. The best quality of the book is not the questions and how I would answer them, but rather the reasoning section that follows each question. If a candidate does not understand the reasoning behind an answer, he or she will certainly falter when the oral board presses for more information. Once a candidate understands the culture of the fire department and **why** the answer is correct, the process becomes a formality, not a stumbling block.

Whichever exam an agency elects to use, it is imperative that you make it through the exam and proceed to the next phase. It doesn't matter how many classes you take or how strong your interview skills are if you don't make it through the process.

The following is an excerpt from my book, <u>Smoke Your Firefighter Written Exam</u>.

Language Expression

1. Choose the answer that is correctly punctuated and a complete sentence.

   ☐ A. I went to the park then to the store.
   ☐ B. She is a very happy person and this makes me smile.
   ☐ C. He is very intelligent, and does well on tests.
   ☐ D. My mother is helpful, and she is a thoughtful individual.
   ☐ E. None are correct.

THE CORRECT ANSWER IS: D

EXPLANATION: This sentence uses a comma to combine two simple sentences.

WHY ARE THE OTHER ANSWER CHOICES INCORRECT?

A. I went to the park then to the store. (It is a run-on sentence that needs punctuation.)

B. She is a very happy person and this makes me smile. (A comma is needed before the conjunction "and.")

C. He is very intelligent, and does well on tests. (There should not be a comma after "intelligent," because the conjunction is not combining two simple sentences.)

Answering Gear Questions

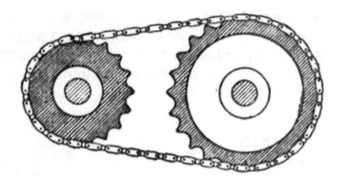

1. If the smaller gear in the above example has 20 teeth, and the larger gear has 40 teeth, the gear ratio would be:

    A.   2:4
    B.   1:2
    C.   2:1
    D.   20:40
    E.   Both B and D are correct

THE CORRECT ANSWER IS: C

EXPLANATION: The ratio of the small gear to the large gear is 2:1. This means the small gear will have to rotate 2 times for each revolution of the large gear. If the small gear had 20 teeth and the large gear had 60 teeth, the ratio would be 3:1. It does not matter what order the gears are in; the ratio is expressed the same. You would further specify whether the gear train was increasing or decreasing by the ratio. An example of how ratios are described would be a gear reduction of 2:1 or a gear increase of 3:1.

Mathematics

82. The firefighters are performing a swift water rescue. Engine 1 is upriver at a bridge and drops into the water a cylume stick in a clear plastic milk carton to check the speed of the river. If Engine 2 is exactly 1.8 miles downstream and sees the illuminated milk carton in 4.2 minutes, what is the speed of the river in mph?

☐   A.   20.2 mph
☐   B.   23.1 mph
☐   C.   25.8 mph
☐   D.   29 mph
☐   E.   None of the above

THE CORRECT ANSWER IS: C

EXPLANATION: Use the formula: Distance = Rate x Time.

Therefore, 1.8 = ? x 4.2 minutes

1.8 divided by 4.2 = rate

= about .43 miles per 1 minute

Lastly, multiply .43 x 60 (minutes in 1 hour)

= 25.8 miles per hour

Interviews

The fire department interview is a unique challenge that is a component of the hiring process for most fire departments. As a general rule, the interview is usually weighted more than any other portion of the exam. It is not uncommon to have each of the other phases of the exam weighted "pass or fail," while the interview is weighted 100% of the candidate's overall score. Simply stated, the interview is the most important phase of the exam process.

Many of the questions put a candidate in a "no win" situation. They are designed to see how the individual can think on his or her feet. While there are often no clear-cut right answers, there are usually automatic fail points.

The best way to learn how to succeed in the interview is to educate yourself on the process. The more you learn about the types of questions that are commonly asked, the more you can do your research, reflect on your own views and attitudes and present appropriate answers.

The competition is so stiff to get a job (usually one hundred applicants for each opening) that fire departments only hire the cream of the crop. One wrong answer will often eliminate a candidate from the process.

Once a candidate understands the interview process and learns what we are looking for, he or she scores well on every future interview. As a result, the candidate will receive multiple job offers.

Since many fire departments only require that a candidate be at least 18 years of age and possess a high school diploma or GED, a candidate theoretically could get hired without having taken a single fire science or EMT course. However, completing EMT training, taking fire science courses and graduating from a basic fire academy will undoubtedly improve a candidate's chances of getting hired.

If a department puts its new recruits through a formal training academy, a candidate who does not possess any of the aforementioned credentials will still have a chance in the hiring process. Other departments require completion of a basic fire academy to even qualify to apply.

The most important thing in the interview process is for the candidate to present him or herself as a person we want to have as a part of our crew.

A candidate can have the most impressive resume, but if he or she is not someone we want to spend a 24-hour shift with, we will not hire him or her to be part of our family. Remember, we have the option of choosing anyone we want. We can train you to be a firefighter; we cannot train you to be a good person.

The best way to improve your interview scores is with practice, or mock interviews. Knock on the door of your local firehouse and enlist the help of the firefighters. They undoubtedly took an interview to get their badge. Some crews will be more current than others on the interview and testing process. Since firefighters are usually not short on opinions, they will probably have a lot to share with you. Listen to what they have to say and incorporate it into your delivery.

Once you have learned the basics of how to take an interview, a private coaching session will certainly enhance your score. I would suggest learning all you can before enlisting the assistance of an interview coach. When you feel you are ready, it is a great investment of your time and money.

The following is an excerpt from my book, Smoke Your Firefighter Interview. Although it may be a review for those who have already read the book, I feel it is important to be exposed to the thought processes behind an interview question.

Tell us about yourself.

My name is Paul Lepore. My family and I live in Dana Point, California. My wife, Marian and I have been married for 16 years and have two daughters, Ashley and Samantha. I grew up in Huntington Beach and spent the majority of my life in northern Orange County before moving south three years ago.

I enjoy sport fishing. My wife and I own a boat on which we spend a lot of time fishing and exploring the waters around Catalina Island. My love of fishing has taken me on some extensive travels through Baja, California. I have even written a book about my passion, called "Sport Fishing in Baja." In addition to the outdoors, I also like playing racquetball and basketball and enjoy riding my bicycle.

I currently work as an electrician. Two years ago I set myself a goal to become a firefighter. Since then I have pursued an education in fire science and have learned all I could about becoming a good firefighter.

Reasoning:

The purpose of this question is to provide you, the candidate, with an opportunity to discuss your personal life. As you may have noticed, I did not mention much about my qualifications. I used this opportunity to talk about my personal life and my hobbies. This kind of question is designed to encourage you to bring out information about your life experiences and personal interests.

Sharing personal information about yourself gives the rater an opportunity to learn what kind of person you are. It also gives the rater a chance to discover something about you that he or she can relate to. That may create a positive feeling, which may result in him/her giving you a higher score. Let me give you an analogy to illustrate my point.

Imagine that our wives work together and have dragged us to their annual office Christmas party. We are sitting at a circular table dressed in our suits and ties. Our wives disappear to mingle with their co-workers. You and I have never met but sense we are in the same boat. Rather than ignore one another, we start talking about such things as where we're from, how many kids we have, where we live, etc. If we have a lot of time to talk, we might even discuss the kind of work we do, how we met our wives, how long we've been married and where we grew up.

Usually when you find a common interest with another person, you tend to want to explore that. For example, if the other person mentions that he likes fishing, I would ask him more about it since I also enjoy fishing. I would mention my interest in both fresh and salt water fishing and encourage him to talk about his fishing adventures.

This example illustrates how common ground can promote conversation, which may then lead into discovering other common areas of interest.

Many candidates mistake this question as an opportunity to outline their resume. This is a serious mistake. The question is designed to

encourage answers about your personal interests. This is your opportunity to show the board who you are. Don't waste time going over your qualifications; rather, use the time to enlighten the board.

By using this opportunity to provide information about where you are from, what you do for fun and any special accomplishments that you are proud of, hopefully someone on the board will identify with something you have said and will feel a connection.

You never know what that connection could be. It may be that they too played high school or college football. Maybe they are from the same part of the country. Perhaps a board member who plays basketball is looking for players for the basketball team. They may have an interest in auto mechanics. It may be possible that you speak a foreign language and your skills may be needed in certain areas of the community. Another benefit of providing personal information about yourself is that once a rater feels a bond with you, he or she is more likely to give you a higher score. It stands to reason that if no connection has been established, you will have to work that much harder for a good score.

Let's say the department has an opening for a seat on the fire engine. They have decided to hire a firefighter to fill the vacancy. Since fire departments are always inundated with prospective candidates when they give an exam, they have the luxury of hiring whomever they want. This wide range of choice makes it more likely that they will hire someone they like.

If you are going to be put straight onto a fire engine, our choices are more limited since prior training is a must. In other words, the department may be looking for someone who has already put him or herself through a basic fire academy at the local junior college.

If we are going to put the new hire through a fire academy, we can hire someone with minimal experience. Firefighters would much rather hire someone who has similar interests, values, goals and morals. I'm not saying they're looking for clones. What they are looking for is someone who fits the profile of a firefighter. They have a much better chance of choosing someone compatible by learning about them personally as well as professionally.

Why do you want to be a firefighter?

Years ago, when I was researching potential career choices, I learned that the father of one of my friends was a firefighter. As I quizzed him about his job, I was struck by how much he loved what he was doing. It was rare to find someone who truly enjoys what he does.

The more I researched the fire service, the more convinced I became that it was the right choice for me. Since then I have visited many fire stations and have gone on several ride-alongs. The reasons I want to become a firefighter are numerous. They include the following:

I enjoy helping people. It gives me great pleasure and it would be very fulfilling to have a profession in which I was able to help people every day.

I would like to be part of a team that solves problems in the community. Whether it is a fire, flood, hazardous material spill, or medical emergency, it feels good to know that citizens can rely on the fire department to help solve their problems.

Being a role model in the community is also important to me. I know children look up to firefighters and I feel we have an obligation to be there for them. I realize the importance of having a smile on my face and being respectful at all times. I also know that firefighters volunteer their time to promote good will within the community. I feel this is a vital part of a firefighter's job. What also appeals to me is the camaraderie that develops in the fire station. Living and working together for 24-hours at a time allows firefighters to develop some incredibly strong bonds.

I like the challenges that a day at the fire station can bring. Even though our on-duty days are planned out, plans can be interrupted at a moment's notice for an emergency response.

Since I am a problem solver, I would thrive on contributing my problem-solving skills to the team. But I know if I'm having difficulty solving a problem, I would be able to rely on the other crewmembers to come up with a solution. The amount of shared knowledge among firefighters is tremendous.

I know being a firefighter will provide many opportunities for learning. There is a tremendous amount of information that a firefighter must learn in order to become competent in his or her job. It would be up to me to set a goal and study hard to achieve that goal. Once I have mastered the roles and responsibilities of a firefighter, I know that I will have many opportunities to test for more challenging roles such as paramedic, engineer, lieutenant or captain.

I like working with my hands. I know the fire service uses a myriad of specialized power, hydraulic and hand tools.

I know the community will always need firefighters. It is comforting to know that firefighters rarely get laid off.

I like the benefits package offered by the fire department. I currently have to pay for healthcare benefits out of my own pocket. I know that healthcare and retirement benefits are part of the fire department's employment benefits package.

The fire department pays good salaries, which will help me provide for my family.

The fire department's flexible schedule would allow me to continue my education and also frees up more time for family activities such as coaching my daughter's soccer team.

I like fighting fire. It is exciting and challenging to arrive on scene and perform hose lays, throw ladders and rescue people. What a great sense of accomplishment that would be.

Since I am interested in medical calls, I would enjoy being an EMT. If the opportunity ever came up, I would like to consider being a paramedic.

Reasoning:

It always amazes me how unprepared candidates are for this basic question. Invariably, when faced with this question, they are usually stumped for an answer. This is the easiest question of all since there is really no right or wrong answer. The panel is trying to determine what your motivation is for wanting to become a firefighter.

Do you believe firefighters have a lot of free time and make good money? If this is your primary motivation, you are in for a rude awakening. If those are your first two answers you are unlikely to get a job in the fire service. If you do manage to get a job with that perception in mind, you will probably have difficulty during your initial training.

These are just a few examples of why candidates want to become firefighters. I suggest you write the reasons that motivate YOU to become a firefighter. When asked the question in an interview, it is important that you not try to remember what you have written down, but rather speak from the heart. If you truly have thought about it, the answer will come naturally. It is discouraging to listen to someone try to figure out the answer to the question during the course of the interview. On the other hand, it is refreshing to listen to a candidate who has given a great deal of thought as to why he or she wants to be a firefighter. Also, try to avoid using "canned" (rehearsed) answers. As a rater, it is discouraging to hear a candidate try to repeat what someone has instructed him or her to say. It is important to speak from the heart, rather than try to parrot some catchy phrase that you learned in an interview class.

Raters usually volunteer to be on the oral boards. As a general rule, most firefighters really enjoy their job. A candidate who demonstrates enthusiasm for the fire service will most likely strike a chord with the raters. If the raters love their job, you can bet they will be looking for firefighters who will also appreciate the job.

Remember, evaluators want to give you a good score. It is up to you to give them a reason to do so.

Notes:

Situational Questions

Situational questions are designed to see how a candidate will respond if faced with adversity. The following are common areas of potential conflict:

1. Moral issues
2. Ethical dilemmas
3. Legal issues
4. Societal obligations
5. Violations of established policies and procedures
6. Interpersonal conflicts

Most candidates focus too hard on what they think the board wants to hear rather than saying what's really on their mind. Here is a typical example:

Moral issue: As you watch the engineer back up the rig, you see him accidentally strike a car. As you approach, you observe him trying to rub out a deep scratch on the vehicle that he hit. When he notices you watching, he tries to make light of it, and tells you not to worry. You tell him that you feel it's important that he bring it to the captain's attention. He explains that it's an "old, junky" car. He adds that in the fire service, everyone sticks together, and asks you to "cover" his back on this one.

Most candidates misinterpret the point of this question. They are confused by the "brotherhood" of the fire service, and believe that firefighters are willing to lie for the good of a co-worker. Nothing can be further from the truth, since a true firefighter doesn't lie, and will refuse to cover anything up.

The best way to answer a situational question like this is to determine how you would handle it in your everyday life. Let's say you and a friend are leaving the parking lot of a restaurant, when he accidentally backs into another car. He leaves a dent, but tells you he doesn't plan on doing anything about it. What would you do and why?

No matter how loyal a friend you are, I don't believe you'd be willing to turn your back on the fact that he just damaged someone's vehicle. You will most likely persuade him to try to locate the owner of the vehicle or call the police.

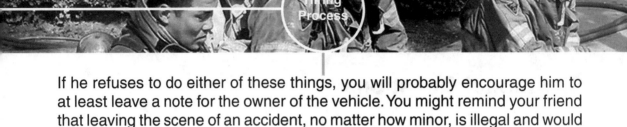
If he refuses to do either of these things, you will probably encourage him to at least leave a note for the owner of the vehicle. You might remind your friend that leaving the scene of an accident, no matter how minor, is illegal and would be considered a hit and run.

For some reason, many candidates believe that firefighters can get away with doing something immoral, unethical and/or illegal.

There are some common rules of thumb when dealing with moral issues. In every situation, it's imperative to do the right thing. In this situation, the right thing would be to step up and take a stand against the action. Remember, even though an action may be legal, it can still be immoral. It's a firefighter's duty to make a moral decision whether it is for himself or a co-worker. A candidate would be expected to know this and respond accordingly.

Firefighters do not operate in "gray" areas. If something is wrong, it is wrong. Even if there's only a perception that it may be wrong, it is usually wrong. Perception often ends up being reality. It is important to maintain the dignity of the fire service. Firefighters are a rare minority of people who the public allows into their homes without a second thought. It is incumbent on all firefighters to protect and honor this privilege.

Ethical questions deal with something that may not be illegal, but either go against society's rules or the cultural rules of the fire department. Ethical dilemmas are often related to violations of departmental policies and procedures. Policies and procedures are often written as a result of either personal injury to a firefighter or civilian or damage to equipment. In almost all cases, they stem from a monetary loss that the fire agency has suffered in the past.

An astute rater will ask a candidate if organizational policies are important. A savvy candidate will undoubtedly nod his or her head yes, and assure him that a firefighter should not, under any circumstances, violate the rules.

If, on the other hand, a candidate believes it's acceptable for a firefighter to violate a policy because it seemed insignificant, it stands to reason that he or she will violate similar policies after becoming a firefighter.

Departmental policies and procedures are meant to be followed. Let's say a rookie firefighter decides that a policy is insignificant and elects to ignore it.

Now let's say an injury or accident occurs as a result of the broken policy. The probationary firefighter will be expected to outline a memo to the chief about the circumstances surrounding the incident. Predictably, the fire chief will want to know why a departmental policy was violated. He will be expected to provide the city manager or board of fire commissioners with an explanation. Imagine the frustration of the fire chief having to explain why a new firefighter violated a policy, and what the consequence of his action will be. Since a probationary employee has no civil service protection and no union representation, a serious infraction could result in his termination.

From an organizational standpoint, you cannot have members follow only the policies and procedures that they feel are important. This would result in an organization that lacks discipline and would eventually collapse. The fire service has adopted many of the military's policies and procedures. This is why the fire department is considered a paramilitary organization. If it's assumed that a soldier would never violate a policy or procedure, why not assume the same with a firefighter?

In most situations, the moral or ethical dilemma wouldn't personally involve you. The dilemma would be for the firefighter who is either asking or implying that you should look the other way. You know what you would do if you were in your fellow firefighter's shoes. You would take the high road and do the right thing. Your challenge will be to convince your comrade to do the right thing. Doing the right thing isn't always easy, but it's the only way to go.

Legal issues are usually pretty clear-cut. Most candidates understand the importance of taking action when a situation is illegal. Candidates who don't understand this will not usually fare well during the interview.

Societal obligations, however, are usually in the "gray area." It can be much more difficult to decide between right and wrong when it involves an action that comes close to crossing the line of good judgment.

In this situation it's important to investigate and gather the facts. If it appears that there has been some type of wrongdoing, you need to make it clear that you would step in and address the situation. The panel does not expect you to suggest there be an in-depth investigation. Your response could simply be that you would address the fact that something was wrong, and would refer the situation to your captain.

Interpersonal conflicts not only create an uncomfortable working environment, but also erode crew unity. There are numerous situational questions that are designed to determine how a candidate deals with these conflicts. While people deal with interpersonal conflicts every day, conflicts in a fire station can be magnified because firefighters live, eat and sleep in close quarters for extended periods of time.

The purpose of these questions is to determine which candidates will get along with others. Candidates who grew up in large families and those who played team sports have an advantage in this area since they are used to dealing with many types of personalities.

When confronted with an interpersonal conflict, it is important to approach the individual and attempt to clear the air. A savvy candidate will suggest asking the other firefighter if he or she is doing something that needs to be changed. Instead of assuming that the other firefighter is off base, it is important to ask (and listen) and then do what you can to improve the situation.

Whatever the cause of the irritation, it is important for the candidate to be humble. As you root your way down to the source of the conflict, it may be that you are not meeting the standard. It may also be that you are not perceived as being a team player.

Situational Question

You see a senior FF put what looks like a department handheld radio in his car. Later, the captain reports that a radio is missing. What do you do?

If you fundamentally believe that firefighters don't steal, this is an easy question. Let me paint a picture for you. I leave my wallet, day runner and all of my valuables on my desk. I don't even have a key to my locker. In any other work environment people would think I was crazy, but in a fire station it is commonplace.

In 20 years of living in a fire station I have never had anything disappear. I have left things in another station and had them arrive in departmental mail (shampoo, toiletry kit and towel). Why would someone take the time to send the shampoo that I left in the shower? It's because the men and women of the fire department are extremely honest.

I trust my life to my co-workers. Everyone who is a firefighter shares the same code of honor. FIREFIGHTERS DON'T STEAL.

The question states that I saw (at least I think I saw) the senior firefighter put a portable radio in his car. It is imperative to gather the facts. The "radio" may have been his or her child's walkie-talkie or a hand held VHF radio that he was charging for an upcoming fishing trip. You just cannot be certain without gathering more information.

I would approach the senior firefighter, already knowing there is a reasonable explanation, because FIREFIGHTERS DON'T STEAL. I would expect him or her to provide me with a legitimate reason. When I get that answer, I would do nothing further. If I can trust him or her with my life, I can certainly trust that same person to tell me the truth about a radio.

If you accuse the firefighter, you are done for. This is the same firefighter who is going to take time out of his or her busy life to make sure that you pass probation, to mentor you and to show you the ropes. You don't want to alienate him or her.

There are a lot of reasons that the crew's portable radio could be missing. These things break on a regular basis. Remember, radios are electronic devices that are carried into harsh environments. They are exposed to heat and water, which are both instant death for a radio. They get dropped on a regular basis. There is a good chance that the engineer tagged the radio and sent it to the shop, and simply forgot to tell the captain. Ironically, you saw the firefighter putting his or her personal VHF radio into the car. OUCH. Imagine how strained your relationship would be after accusing your senior firefighter.

I keep stating that firefighters don't steal. I truly believe this to be true. Is there a one in a million chance that the firefighter stole the radio? Yes, there is. That's why I gathered the facts. I approached (not confronted, as this implies a fight) and asked the senior firefighter about the radio. When he or she gave me a reasonable explanation, I was done with the matter.

If the firefighter had told me to mind my own business, take a hike or keep my mouth shut, we would now have a problem. I am not saying the firefighter is stealing, but I am suspicious.

I would explain that I was not sure what was going on here (you still don't want to accuse him or her of stealing) but I was really uncomfortable with the situation. I would state that we need to take this matter to the captain.

The firefighter tells you to mind your own business.

I would explain to him or her that I have an obligation to the department, the citizens and myself to bring this to the captain.

The firefighter tells you that if you bring this to the captain's attention, everyone will think you are a snitch.

Unfortunately that is a chance I will have to take. Again, I have an obligation.

I would then explain to the firefighter that he or she would look better in the eyes of the captain by coming forward on his or her own. If the firefighter is unwilling to go to the captain, I will offer to accompany him or her, or go to the captain on my own if necessary.

Notes:

_____

_____

_____

_____

_____

_____

_____

_____

The Chief's Interview

Alan Patalano is a Deputy Chief for the Long Beach Fire Department in Southern California. He has agreed to share his thoughts and ideas on what he is looking for from a candidate during a chief's interview.

There are dozens of people around who will be glad to offer advice on how a candidate should perform during a Chief's interview. They will tell you about the theory of interview questions, body language, dress and presentation. I don't have expertise in any of those areas; instead, what I have is the experience of conducting Chief's interviews from the perspective of a Chief Officer and from the perspective of sitting in the room after the interviews are completed and actually deciding which candidate gets a job offer.

The Chief's interview is far different from the structured oral interview that you may take during the initial testing phase. The initial interview usually asks every candidate the exact same questions, in the exact same order. This is done so that the exam is consistent for everyone. The Chief's interview does not operate in this fashion. In the Chief's interview I am free to ask questions of each candidate based on his or her resume, experience, education, background and responses to previous questions. I do not need to ask each candidate the same questions. This is an important point. My questions are based, in a large part, on your responses to prior questions.

I evaluate your responses in several ways, including:

1. How well do you communicate?
2. Are your answers thought out?

3. Are you confident?

4. Are you truthful?

Let's look at each aspect:

How well you communicate has a huge impact on your overall score. Your ability to utilize the spoken word to convey a message or make a point is the foundation of a great score. The first portion of good communication is listening. What do I mean? Simply stated, to develop a great answer you must know what question I am asking. It is not uncommon to stop a candidate a couple of minutes into a great answer because he or she is answering the wrong question! There are several reasons why this happens:

- The candidate anticipates particular questions before arriving at the interview, classifies the question as one of his or her preconceived questions and provides the answer.

- The candidate has a list of predetermined answers and utilizes the canned answer that is closest to the question I ask.

- The candidate formulates a reply without listening to the complete question.

- The candidate is nervous and gets off-track while answering.

So before you can develop a great answer you should listen carefully to the question in its entirety. If you are unsure of what is being asked, then ask for the question to be repeated and/or clarified. This not only allows you to provide the best possible answer, but also shows that you are not afraid to speak up when needed to avoid mistakes (a good quality to have on the fire ground). But do not make it a habit to ask for every question to be repeated. This might only show that you are not attentive.

Once you determine what the question is, make sure you take the time to formulate a great reply. Many times I no sooner finish the question before the candidate starts talking. I always think to myself, "I wonder if the candidate was listening when I was talking."

Tone of voice, volume and grammar all impact how I perceive your answer. An angry or aggressive tone makes me question how you may respond to the public during emergencies, especially when you are under stress. Low volume indicates a candidate may be timid or lack confidence. Poor grammar or slang makes me question your maturity. Remember that good communication is predicated on providing information in a format so that the **listener** (i.e. the interview panel) can understand it and not on the way you like to say it.

Next I like to see that your answers are thought out, logical and realistic. Once I ask the question, you should be able to walk me through the sequence of events or the steps you would take. As an example, if the question asks about your education, your response shouldn't start with high school, then discuss grade school, then a course you are currently taking and then your college experience. It should be presented in a logical sequence: grade school, high school, college and the current course. It is confusing to the interviewer when the answer is presented in a disorganized fashion and makes me wonder if everything you do is disorganized.

Answers also have to be realistic. If asked a situational question about which task you would perform: 1) pull a hose line to a door, 2) hook to a hydrant, or 3) raise a ground ladder, the worst answer would be, "I would do them all because I am young and strong." It's not practical and shows a lack of understanding of the real world. On the fire ground we are faced with choices and every firefighter must be able to analyze facts and make decisions. I expect to see this same quality during the interview.

Another big quality I look for during the interview is how you represent yourself. Do you appear confident? Are you sure of yourself? Your answers should reflect your confidence in your skills and abilities. An answer that is vague or noncommittal demonstrates a lack of confidence. The nature of our business makes confidence during emergencies a vital personal quality. Can you make a decision and then act on it? Needless to say, there is no crying during the interview!

Finally, do not let me catch you telling a lie, stretching the truth or telling only half the story. I am willing to overlook past behavior (up to a point) if you have shown that you have changed that behavior. I won't consider it past behavior if I find you to be dishonest or unwilling to share all of the facts during the interview. That is your current behavior and is unacceptable. It will not matter to me if you can offer a good excuse for why you weren't honest initially because I will already be looking for a better candidate. I cannot stress this enough. If I catch you in a lie you will not get a job offer today or for the life of the list, period.

It is very important to understand that during the interview I am looking for candidates who will be able to work with my firefighters for 30 years. We can train you to pull hose, take a blood pressure and operate a hydraulic rescue tool. What we can't train you to do is act in an honest, ethical manner or be professional or compassionate. You must have those traits "built-in" before you arrive for the first day of drill school, so I look for those qualities during the interview.

Education shows that you can commit to a course of action and follow through until completion. Work history shows loyalty and commitment. Community activities show that you believe yourself to be part of something greater than just yourself, your family and friends. How you dress shows that you consider yourself important and respect the job and those who perform it. All of these things serve to assist me in "seeing" the real you. No single fact, statement, or resume line assures you a job offer. Instead it is a compilation of all of your various education, background, experiences and presentation that helps you to rise above the other candidates and secure a position.

I have offered positions to candidates with years of firefighter experience and to those without any experience at all, to those with extensive education and to those with only a GED, to candidates with a list of certificates and to those who didn't have a single piece of paper except what was required to apply. What they all had in common was desire, commitment, honesty, loyalty, compassion and a dedication to serving a greater good. If you possess these qualities and can demonstrate them to me during an interview, then there is a very good chance that within a year I will be shaking your hand and welcoming you as the newest member of my department.

Resumes

Many firefighter candidates believe that their resume doesn't have to be perfect. Some feel as if they can explain themselves in the interview: "My resume doesn't have to be perfect; I'll worry about it if they ask me."

In most testing processes, a resume is submitted with the initial application. By the time a candidate reaches the oral interview, the panel has already reviewed the application and attached resume. If there are any discrepancies, the evaluators will circle them and ask the candidate during the interview. Inconsistencies can include a typographical error, a gap in employment or a lack of follow through in completing a degree. Whichever the case, your resume is a statement of who you are that arrives long before you do.

Oral interviews are usually scheduled in rapid succession. As a result, the interviewers do not have a lot of time to extract information from the resume. Studies show that the average interviewer spends 8-15 seconds reviewing a resume. If your resume is not clear and concise, it will be of no value since the reader will not dig deeply for information.

Standard fire department resumes should be kept to one page. Some departments will allow two pages, but these seem to be in the minority. Consequently, the resume must be clear and concise and spell out your accomplishments. If all of your "gold" is buried, the board will not discover it. Before putting down a single word, take a few moments to outline your accomplishments. Decide what your best accomplishments are and build around them.

Many of the candidates we encounter are working professionals who have decided on a career change. The resume that worked to get them a job as a teacher, computer programmer or stockbroker does not work for the fire department. Fire departments are not looking for resumes loaded with "credentials" that do not apply to the job. We are not impressed if you are proficient on Java Script or Excel spread sheets. Are these helpful? Well, technically yes, but we would rather see EMT, fire science courses and practical hands on experience.

Common mistakes these professionals make is taking up too much space on the descriptions of their job duties. We really are not interested in the fact that you were part of a technical team which incorporated new software into the computer network or that you were in charge of adding the food coloring to the drink dispenser at the fast food restaurant. My suggestion is to keep the job descriptions brief. If we want to know more, we will ask you. Use the valuable space listing items related to the job of a firefighter.

Resumes should be created using clear, concise language describing tangible, no-nonsense skills: competent in Vertical Ventilation, able to converse in Spanish…. Always steer clear of fluff words and phrases like the following: self-motivated, excellent track record and honest. That's up to the panel to decide.

Another thing to avoid putting on a fire department resume is "References available upon request." If you are going to be selected to continue in the employment process, you will be asked to provide them. Besides, nobody ever puts anyone down as a reference who would say something negative about him or her. In other words, references really don't add much to the package.

Email addresses aren't worth the space they take on the resume. This is particularly true if your personal email address is unconventional or "cute." Leave them off the page. Besides, email addresses change like the weather and your application will be on file for the life of the list, which often exceeds four years. If the fire department needs to contact you, they have your phone number and address.

If you are like many of the applicants who are trying to "squeeze" in all of their related coursework and certificates, you may try organizing your achievements in two or three columns instead of listing them in a single column. This allows you to get the maximum usage out of each line.

Evaluate the relevancy of your early experience. Taking a first aid and CPR course is great, but it loses its importance once you have completed an EMT course. In fact, if it is still in place the reader gets the impression you are trying to "fluff" up your resume.

A clean, clear and concise resume can be designed on your personal computer using a resume program. Once the basic template has been designed, the user is able to add or delete information with ease. In addition, the resume may be customized to include the name of the department you are applying for.

If you are having difficulty designing your resume, enlist the services of a friend. If this is not possible, professional resume writers can be located in the phone book. The downside of having your resume professionally done is that each time you want to add a new class or accomplishment, you have to pay for it. It is also costly to tailor it each time you apply for a different department. A solution may be to have a professional resume writer create a template and provide you with a disc that you can save in your computer. This would allow you to make future modifications without incurring further expense.

Always proofread your resume. Once you are finished with it, give it to a friend or two. It is very difficult to proofread your own material. You know what it should say and overlook the errors.

I have included a few sample resumes to give you some ideas. Feel free to use any of them as a template for getting started.

Notes:

Sample Resumes

Robert Jones
229 Turquoise Drive, Morro Bay, CA 93420
Phone: (909) 281-1019

OBJECTIVE: *Career Firefighter/Paramedic for Douglas County Fire Department*

PROFESSIONAL EXPERIENCE

PARAMEDIC – SAN LUIS AMBULANCE SERVICE, San Luis Obispo County · · · · · · · · · · 2003 to Present

Responsible for patient assessment, care, packaging and advanced life support through the response to emergency calls; operation/inspection of ambulance, radio and emergency equipment.

FIREFIGHTER – ARROYO GRANDE FIRE DEPARTMENT, Arroyo Grande, CA · · · · · · · · 2001 to Present

Responsible for the protection of life, environment and property through the suppression and prevention of fires and in the emergency response to calls; patient assessment, care and packaging; weed abatement and public education.

EMT – SAN LUIS AMBULANCE SERVICE/AMR, Central Coast CA · · · · · · · · · · 2002 to 2003

Responsible for patient assessment, care and packaging through the response to emergency calls; operation/inspection of ambulance, radio and emergency equipment.

OWNER/OPERATOR – BABY PROOFING COMPANY · · · · · · · · · · 2001 to 2002

Service and installation of baby proofing safety products.

GENERAL MANAGER – STAPLES, San Luis Obispo, CA · · · · · · · · · · 2000 to 2002

Responsible for the successful opening and operation of a new store location with above projection sales.

MANAGER – TARGET · · · · · · · · · · 1998 to 2000

Ensured all team members kept customer service a number one priority.

STORE MANAGER – RADIO SHACK · · · · · · · · · · 1991 to 1997

Developed outstanding sales team to maximize sales over $1 million.

DESIGN DRAFTER – FLEETWOOD ENTERPRISES, Riverside, CA · · · · · · · · · · 1989 to 1991

Create, maintain and update floor plan, interiors and framing drawings in the motor home division.

EDUCATION AND PROFESSIONAL TRAINING

A/S degree Fire Science (Spring 2004)	EMS Academy
Firefighter 1	ICS 100/200
Technical degree Architectural Drafting	BLSC
EMT-D	Hazmat FRO
Hazmat Decon	Rescue Systems 1
Ambulance Drivers Certificate	S-190/290
CEVO II	NHTSA Child Passenger Safety Technician
S-205	ACLS
BTLS	PEPP
PALS	WMD Response
EVOC	

Samuel Adams
1234 Vista Royale Drive • Dana Point, California 92867
(714) 555-7855

OBJECTIVE: Appointment to a firefighter position utilizing safety, leadership and public relation skills with the Orange Fire Department.

EXPERIENCE:

Huntington Beach Fire Department   Summer '03 to Present
Ambulance Operator/Fire Intern
Responsibilities include emergency medical calls and treatment, providing paramedic assistance and ensuring safe, rapid transport to emergency facilities. Accountable for patient billing and pre-hospital reports, working knowledge of all EMT equipment, apparatus, effective mapping skills, station and vehicle maintenance.

Central Net Training Center   Summer '03 to Present
Student Intern
Assist with setup and training with flashover classes. Responsible for facility and equipment maintenance cleaning the facility and stocking supplies. Complete various other duties as assigned by Battalion Chief.
·   *Under direct supervision of Al Marland*

Santa Ana College Basic Fire Academy   Spring of 2003
Academy Recruit Chief
Directly managing, leading and teaching recruits. Interacted with top-level academy officials, chiefs and recruits; disciplined academy members on a case-by-case basis; demonstrated excellent team-building skills.
·   *7th Academy recipient of the Captain Donald Michael Hibbard Inspirational Award.*

State of California – Huntington Beach & Bolsa Chica State Beach   2001 to Present
State Lifeguard
Aquatic safety, medical aids, teamwork, equipment maintenance, physical training, code enforcement and drowning prevention. Selected to lifeguard at CALPALs beach play day.
·   *Rookie of the Year recipient from Huntington State Beach Lifeguard Association.*

California Pizza Kitchen, Inc. (Santa Ana)   1999 to 2002
Host/Food Server
Typical server duties (extensive interaction with public), opening and closing store, host of the facility, responded to and resolved customer inquiries and complaints.

Orange County Fire Authority Explorer Post 9634   1997 to Present
Fire Explorer, Post Captain and Post Advisor
Coordinating post events, teaching/training, planning and administering drill nights and assumed the role of the "key youth leader" of the post. Chosen to speak at the 2000 Spurgeon Award's Ceremony.
·   *Recipient of Young American Award by Boy Scouts of America (Orange County)*
·   *Academy Chief at the Orange County Fire Exploring Academy*

EDUCATION SUMMARY:

Santa Ana College (Santa Ana) / Santiago Canyon College (Orange)   1999 to Present
Course study – AS degree in Fire Technology (With Departmental Honors). Related courses taken include: hearing impairment communication. Member of Phi Theta Kappa International Honors Society.

Villa Park High School (Villa Park)   1995 to 1999
Courses included two R.O.P. classes related to Fire Technology program, one year of photography. Participated in wrestling and football (3-year varsity letterman).

PERSONAL SKILLS:

Physical fitness: Surfing, running, swimming, bicycling and weight lifting.

Skills: Sign language

4163 E. Main Street   Phone (805) 989-1393
Simi Valley, CA 93063

Robert Jones

Objective

To serve as a Firefighter Paramedic on a team of dedicated professionals with a goal of demonstrating excellence in a cooperative environment.

Education

Graduated May 2004   UCLA Center for Prehospital Care   Los Angeles, CA
UCLA – Daniel Freeman Paramedic Education

Graduated May 2002   San Diego State University   San Diego, CA
Bachelor of Liberal Arts and Sciences in Communicative Disorders

Work Experience

June 2003 to Feb. 2004 American Medical Response   North Hollywood, CA
Emergency Medical Technician, Associate Field Training Officer
Conduct field training, education and evaluation of new employees.
Provide exemplary patient care and customer service.
Utilize medical equipment to carry out basic life support functions.
Ambulance operation.

Jan. 2003 to June 2003   American Medical Response North Hollywood, CA
June 2002 to Dec. 2002   American Medical Response   San Diego, CA
Emergency Medical Technician
Patient assessment and care.
Utilize medical equipment to carry out basic life support functions.
Ambulance operation.

June 2000 to June 2002   Wells Fargo Bank   La Mesa, CA
Lead Teller
Train and supervise new tellers on transactions, bank policy and procedure.
Review and approve teller transactions for tellers with less authority.
Process common and complex teller transactions.

Accreditations and Licenses

NREMT-P, ACLS, PALS, PHTLS, Firefighter 1 Academy, HAZMAT First Responder Operational, Incident Command Systems 200 / Basic ICS, 4 classes away from obtaining Fire Technology Certificate.

Additional Skills

Working knowledge of automotive, carpentry, masonry, painting, plumbing and tile skills.

Languages

American Sign Language

Robert Jones
28283 Via Fuego   Laguna Niguel, CA 92677   (949) 215-7887

WORK EXPERIENCE

1/04 – Present   **EMT**
<u>Doctor's Ambulance</u> – Laguna Hills, California
·   Assist paramedics with patient care, BLS, Transport

6/93 – Present   **Vice President/Art Consultant**
<u>Miranda Galleries</u> – Laguna Beach, California & Aspen, Colorado
·   Supervised training sessions, Worked directly with artists, Sales, Administration, Packing and Shipping

EDUCATION

2/02 – 11/03   <u>Santa Ana College</u> – Santa Ana, California
AS degree: Fire Technology – Public Fire Service
Certificate: Fire Technology – Public Fire Service

3/03 – 6/03   <u>128th Basic Fire Academy</u>

* EMT-B / National Registry	* Firecontrol 3
* CPR / American Heart Association	* ICS–I200
* HazMat FRO	* Rescue Systems 1
* S-190 Wildland	* Auto Extrication

1988 – 1993   <u>University of California Los Angeles</u> – Los Angeles, California
Five years undergraduate
Degree: Bachelors
Major: History

1984 – 1988   <u>Laguna Beach High School</u> – Laguna Beach, California

SPECIAL SKILLS

·   Bilingual in Spanish
·   Bilingual in Portuguese

VOLUNTEER WORK

·   <u>Meals on Wheels</u> – Home delivery of lunches to elderly
·   <u>Special Olympics Pancake Breakfast</u> – Fullerton Fire Department

Robert Jones

3649 Seagull Drive Long Beach, CA 90808
Phone: (562) 421-5154

OBJECTIVE

To obtain a full time Firefighter position with the City of Long Beach and to begin a life long career in the fire service.

QUALIFICATIONS

I am presently a cadet captain in the La Habra Heights 35th Fire Academy. November 2, 2003 I completed the Biddle Validated Physical Ability Test in 6:52 at Central Net Training Facility in Huntington Beach. I have a working knowledge of the fire service. I have a solid understanding of the basics of building construction, plumbing and electrical. Member of Long Beach Search and Rescue, 1986-88.

EDUCATION

2004 – Present	La Habra Heights 35th Fire Academy, Cadet Captain
2003 – Present	Long Beach City College, Fire Technology 10.5 units
2004	University of Phoenix Bachelors degree, Business Management
1998	Long Beach City College Associates degree, Business Management

EMPLOYMENT

2004 — Dispatcher/Reserve Fire Inspector, La Habra Heights Fire Department

1999 – Present — Terminal Manager, Pelican Marine

1. Work with Customs and Border Patrol to ensure Homeland Security procedures are in place and meeting specified requirements.
2. Work with union officers and representatives from six different labor unions to resolve safety and contractual issues.

1998 — Vessel Superintendent, SSA Marine

1. Ensure that a safe work environment is maintained and that all employees are working in a safe and productive manner.

INTERESTS

Include but are not limited to, activities with my wife of 11 years and my three daughters (8, 4, & 1 yrs old), outdoor sports, helping coach my daughter's softball team, restoring my 1953 Chevrolet truck and home improvement projects such as bathroom remodels, door and window installations and installation of ceramic tile.

Physical Fitness

By Marian Lepore

Marian Lepore is a Physical Therapist working in Southern California. She has worked with clients of all ages, emphasizing both fitness and wellness, and rehabilitation post injury.

Physical fitness is a critical part of a candidate's preparation for the fire service. It takes a great deal of effort and commitment to achieve the level of fitness required to pass the Physical Ability Test (PAT), complete the fire academy and, ultimately, safely and effectively perform the job. As a firefighter, your coworkers and the community stake their lives on your ability to perform physically, whether it is to pull hose, rescue a crewmember in full gear, or throw a heavy ladder.

Fitness has four major components: aerobic capacity, muscular strength, flexibility and body composition. Total fitness requires a total body approach. The overall level of physical fitness required for the fire service does not happen overnight. It takes time to build up the necessary aerobic capacity, muscle strength and endurance. Whether you are a highly trained athlete or are just beginning your physical training, it is important to develop a training regimen that will prepare you specifically for the challenges of the PAT and the fire academy.

Everyone's physical ability is different, based on genetics, gender and conditioning. Some people are naturally more muscular, while others never seem to be able to bulk up. Some enjoy running marathons, yet others avoid climbing stairs whenever possible. The good news is that everyone's muscles respond the same way – they grow stronger in response to adequate resistance.

Start your exercise program at a relatively easy level, and gradually increase the intensity and duration. Check with your doctor if you have a history of a medical condition or experience difficulty with your exercise program.

Anatomy and Physiology

Your body uses two energy systems to deliver oxygen to working muscles: the aerobic and the anaerobic systems. During aerobic activity (e.g. jogging), the cardiopulmonary system is able to deliver adequate oxygen to working muscles without building up waste products. In anaerobic activity, the exercise intensity is so high that the cardiopulmonary system cannot keep up with the muscles' demand for oxygen. The body uses the glycogen (carbohydrates) stored in the muscles to fuel activity without the use of oxygen. Anaerobic activities are intense but of short duration (e.g. sprinting).

Muscle cells are composed of myofibrils, which are bunched together into fibers, and in turn grouped into muscles. Muscle fibers are either fast-twitch or slow-twitch. The fast-twitch muscle fibers assist more with anaerobic activities requiring power (a great deal of strength over a short period of time), while the slow-twitch fibers are more involved with aerobic activities (prolonged activity).

Since the same muscle groups may be recruited for each type of activity, training for both strength and endurance can produce a confused physiological state. However, if you carefully time your sessions to perform strength and endurance training on different days, you can avoid this interference effect.

Cardiovascular Endurance Training

Cardiovascular endurance is the ability to perform low intensity exercise for long periods of time without undue fatigue. Endurance training results in an improved ability to transport and utilize oxygen. The anaerobic threshold is also increased, so the body can exercise at a higher workload without fatigue. Other benefits include a lower heart rate and blood pressure both at rest and during daily physical activities.

Exercises that greatly increase the blood flow to working muscles for an extended time improve cardiovascular fitness and endurance. These activities include running, bicycling, swimming, skating, hiking and active sports that use large muscle groups. It is generally recommended that the target heart rate for cardiovascular training is the following formula: 220 – age x 60%-80%. Although this standardized formula cannot take into consideration each person's individual fitness level or capability, it is a starting point.

Strength Training

Cardiovascular training does not increase the size of muscles. Muscle mass is increased only through frequent muscle contractions against a high resistance. (e.g. weight training). Individual muscle fibers become thicker, thus increasing muscle size. When a muscle is overloaded, it will adapt by increasing in strength.

The One Repetition Maximum (1RM), a measure of muscle strength, is the maximum resistance that a muscle or muscle group can overcome through the full range of motion in a single maximal effort. Contractions of at least 80% of 1RM are ideal for maximal strength development.

Muscular endurance is the ability to repeat muscle contractions multiple times without fatiguing. While strength is increased by using moderate to high resistance with few repetitions, muscle endurance is improved by using low resistance with high repetitions.

Strength is increased as specific groups of muscles are exercised against progressive resistance. The progressions must be gradual to allow the muscles time to recover and adapt. Muscle recovery time is necessary to avoid overuse and inflammatory changes.

Find the balance between stressing muscle fibers enough to stimulate muscle growth, and lifting too much, which can cause injury. Develop a strength training program that addresses the major muscle groups of the body. Emphasize a total body approach with a gradual progression of resistance, duration and frequency. Consult a professional trainer if needed.

There are three types of muscle contraction: isometric, concentric, and eccentric. Isometric (holding) contractions do not contribute to muscle endurance, speed of movement, or motor control, and will not significantly increase strength during tasks involving movement. Concentric (shortening)

and eccentric (muscle lengthening under tension) contractions will both result in increased muscle strength when performed at higher intensity with few repetitions, or increased muscle endurance at lower intensity with more repetitions. Although eccentric contractions have been found to produce larger increases in muscle strength, they also result in greater post-exercise muscle soreness.

Both concentric and eccentric strengthening are important. Strength improvements carryover most effectively when exercises closely match the functional task you are trying to improve.

Delayed Onset Muscle Soreness (DOMS) is a common companion to strength training. The higher the intensity of exercise, the worse the DOMS. The muscle soreness usually peaks within 48 hours, and lasts four to six days.

Possible causes of DOMS include muscle spasm, muscle fiber or connective tissue disruption, and edema. Severe DOMS will interfere with your training routine and decrease performance. You can reduce DOMS occurrence with proper warm-up and cool down, stretching and a more gradual increase in resistance training.

Stretching

Flexibility training, or stretching, is an important part of fitness. It keeps muscles supple, prepares the body for activity, and helps prevent injury. Stretching is most effective after muscles are warmed up.

In order to stretch effectively, follow these basic rules:

- Relax and stretch slowly, without bouncing.
- Stretch to the point of muscle tension, not pain.
- Breathe regularly; do not hold your breath.
- Hold the stretch for 20-30 seconds. Your muscle will initially tighten, then relax as it fatigues.
- As the muscle relaxes, increase the stretch slightly as you exhale. Hold for 30 seconds.

- Stretch regularly to achieve a permanent increase in range of motion (ROM).
- Avoid overstretching, as it can increase the risk of joint injury.

Stretch a target muscle by isolating it. In other words, make sure it is the primary muscle being stretched. For example, to stretch your hamstrings, do not stand and touch your toes. This technique would stretch your lower back along with both hamstrings. One proper technique is to stand with one foot slightly in front of the other, and bend forward at the hips. Keep your knees straight and your back extended. Place your hands on your rear leg or on a nearby surface to support your back. You will feel the stretch in your front leg.

Strenuous exercise promotes muscle tightness and inflexibility. The goals of stretching are to achieve and maintain a normal ROM in all the major joints, and to obtain correct postural alignment of the spine. Flexibility can improve fitness by reducing muscle tension and risk of injury, while promoting relaxation, improved posture and ease of movement. Consult a professional trainer or physical therapist to help you develop a stretching routine with proper technique and goals.

Body Composition

Fat is an essential part of your body and is necessary to function normally. Excess body fat, however, puts unnecessary stress on your body, slows you down, and puts you at higher risk for a multitude of diseases.

Body composition is the ratio between body fat and lean body mass (muscles and bone). You cannot determine body composition simply by weighing yourself. Muscle weighs more than fat, so two people who weigh the same could have completely different fat-to-lean mass ratios.

A health or fitness professional can easily determine your body composition by taking fat measurements at certain places on your body. The ideal body fat level is usually considered to be 10-15% for males, and 15-22% for females. Don't be concerned if you gain weight during your fitness program. Your weight may increase as you replace fat with muscle mass.

Exercise Tips

Make your training routine part of your daily schedule.

- Don't procrastinate.
- Find activities that you enjoy and are convenient for you.
- Incorporate running and stair training into your routine.
- Make fitness a way of life.
- Use the buddy system. Enlist the help of family, friends or a personal trainer.

Exercise frequency will vary according to the intensity and duration of the activity. Standard frequencies are three to five times per week for endurance training, and two to three times per week for strength training. Find the training regimen that meets your personal needs.

Maintain muscle balance. Don't focus on only one set of muscles or one part of the body. For example, if you strengthen a certain muscle, but don't strengthen the opposing muscle, you are setting yourself up for injury as the unopposed muscle shortens and the joints involved become less stable and supported.

Make sure that repetitions are slow and controlled, so that you rely on muscle rather than momentum.

Invest in proper footwear that provides adequate shock absorbency, supports the natural arches of the feet, and is designed for the kind of activity you do. The wrong shoe selection can result in painful injuries in the legs or back.

Sweating is your body's natural cooling mechanism. It is the evaporation of sweat from the skin that produces the cooling effect. When you exercise, wear clothes that are comfortable and allow evaporation of moisture, such as natural fiber (cotton) clothing.

Do not hold your breath while exercising. It can adversely affect your blood pressure and heart rate.

Begin your exercise routine with at least a five-minute warm-up to increase circulation, breathing and body temperature. This will decrease the risk of injury and prepare your body for the workout.

End your exercise session with a five to ten minute cool down (walking, light exercise) to allow your body to adjust and recover.

Adequate nutrition and hydration will help improve the effectiveness of your training routine. Drink fluids before and during exercise to maintain hydration. Replenish fluid and muscle glycogen stores soon after exercise.

Allow adequate recovery time between workouts. If you find that you are constantly fatigued, sore, injured or just not increasing your strength level, you may be over-training. You must find a balance between waiting long enough for muscle recovery and waiting so long that you lose the benefits of training.

Injuries

Large muscles do not protect you from injury. Some of the strongest and most physically fit firefighters have been out on medical leave due to back injuries or other musculoskeletal problems. Improper lifting technique, poor posture and tight muscles (decreased flexibility) may put you at greater risk for injury.

Pain is your body's way of telling you that something is wrong. If pain is excessive or prolonged, don't try to tough it out and work through the pain. A change in your footwear, exercise progression, lifting technique, stretching routine or rest schedule may be all that is needed. Seek the assistance of your physician, physical therapist or other qualified medical professional to take care of an acute problem before it becomes chronic.

Specificity of Training

The power of a muscle contraction is a combination of both muscle fiber strength and facilitation (nerve stimulation). An increase in strength does not always carry over to an increase in function. Improvements in function can be seen as coordination, movement control and muscle recruitment improve through practice. It's necessary to train for a specific task. For example, if you want to learn to pitch a fastball faster, practice at a low load for high repetitions.

Training is most effective when the exercise resembles the specific task in terms of sequence and speed of movement, muscle recruitment, and balance

and coordination demands. Training in a specific activity will improve skill and coordination and decrease the oxygen cost of that particular task.

The Fire Department PAT will challenge your aerobic and anaerobic capacities, upper and lower body strength and endurance, grip strength, agility, balance and proprioception (awareness of body positioning). By using the specific tasks of the PAT as part of your exercise routine, you will maximize your performance improvement. Without specificity of training, improvements in strength and endurance will be wasted.

This section on physical fitness is simply an overview. It cannot take into account each candidate's individual knowledge, fitness level and genetic predisposition for strength and endurance capacity. Seek the guidance of a professional trainer or other qualified individual to develop a safe and effective training routine that will adequately prepare you for the PAT.

Notes:

Preparing for the Physical Ability Test

By Jerry Graham

Jerry Graham is a Firefighter/Paramedic for the City of Orange Fire Department in California. His personal commitment to fitness puts many much younger firefighters to shame.

The CPAT and Biddle are examples of the Firefighter's Physical Ability Test (PAT), which assesses the ability to do the functions of a Firefighter in a timed and graded sequence. These tests are among the most common exams being administered and have held up in the court system for their fairness and applicable standards, which measure and mimic the various duties of a Firefighter. Many departments elect to use their own PAT.

I am going to insert the "Don't engage in any strenuous physically challenging activities unless you have been medically cleared by your physician" right up front. There are many medical conditions aggravated by physically demanding tests. If you are at high risk, do everyone a favor and get cleared prior to the event instead of dropping like a sack of rocks at the test.

The Truth Hurts

There is no one "silver bullet" available to get the best score in any of the areas of testing for the position of Firefighter. By a "silver bullet" I mean one text to study, one fitness routine, one oral interview response that will get you the badge 100% of the time. The process comes down to this: "You have a lifetime to prepare for most, if not all, aspects of this and many other jobs."

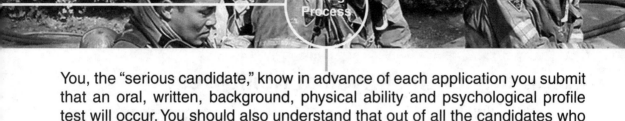

You, the "serious candidate," know in advance of each application you submit that an oral, written, background, physical ability and psychological profile test will occur. You should also understand that out of all the candidates who apply, only *10%* of them are s*erious.*

The *Competitors* (the serious competition), as I will refer to them, have the mind set that pursuing the greatest job in the world (that of the firefighter) requires that they be ready at all times. They don't always know who is going to test and when, thus the "lifestyle" of the competitor allows them to be prepared with very little notice.

A few augmentations to the resume, the timely return of the perfectly typed application, and they are good to go. The "lifestyle" becomes the "life routine" and makes last minute preparation unnecessary. This applies to the promotional candidate also.

It is very important to understand this concept, because in the long run it is this willingness to sacrifice and acquire the mental edge that will make a successful candidate. I will now step off the soapbox.

Let's Get to the Point

Imagine for a moment that you are going to ask your body right this minute to go out to the track and run an 800-meter (two ¼ mile laps sprinting as fast as your legs will allow) event for time, and expect to be in the top 10% out of 800-1500 runners. Is that a reasonable expectation? Proper physical and mental conditioning is key to stacking the odds in your favor. By seeking out the types of tests your target department might administer, and taking every opportunity to practice them, it gives the mind a familiar comfort. If you have previously done these tasks and know how your body will react to the stress, you have a definite advantage.

Fuel Your Fitness

My philosophy includes having a way of life that incorporates physical conditioning as a lifestyle, not a burden or chore.

Some examples of the lifestyle will focus on a Nutrition Plan. "You are what you eat" was the mantra of years past, but to this day it can pay huge benefits if applied properly. Correct fueling of the machine will carry the *competitor* through successful completion of the tasks. Proper food intake and rest are key components.

It takes energy to perform the exercise required to build adequate strength and endurance for the physical ability test. The energy should come from nutritious foods that can help sustain blood sugar levels throughout the day, not from sweets that only give a quick energy boost. Fuel your workouts with a pre-exercise snack to prevent symptoms of hypoglycemia (i.e. fatigue, lightheadedness, weakness) and to build up glycogen (carbohydrate) storage in your muscles. After exercise, refuel your muscles with carbohydrates within one hour to rebuild the glycogen stores.

Experiment with different foods and fluids to determine the best choices for you, when and how much to eat or drink. Your goal should be to maintain adequate physical stamina and muscle fuel throughout your workout without experiencing an upset stomach or symptoms of hypoglycemia or dehydration. Make sure you have determined what foods and fluids work best for you *before* the day of the PAT.

If your focus is on losing weight through exercise, it is still important to fuel your workouts. Although it is true that you will burn more fat by not eating before workouts, your workouts will be less intense and of shorter duration. Body fat is lost by creating a calorie deficit each day. Harder workouts will burn more calories and can help contribute to this calorie deficit.

If your focus is on increasing weight, you must eat more calories than your body needs to maintain its weight. It's important that you add on lean and healthy pounds, not fat pounds. Increase your calories gradually to achieve a slow and steady weight gain. In theory, consuming an extra 500 calories each day will add one pound of weight each week.

> My philosophy includes having a way of life that incorporates physical conditioning as a lifestyle, not a burden or chore.

Most people believe that it is essential to eat a lot more protein in order to build muscles. Actually, the typical American diet usually contains enough protein for muscle building. When you are strength training, it's more important

to eat additional healthy carbohydrates to fuel your workouts than to change to a protein-heavy diet. Carbohydrates should supply 60-65% of the calories you eat each day, protein should supply 12-15%, and fats no more than 25-30%. If you have questions about your diet, consult with a sports nutritionist or registered dietitian.

Seasonal Training

The time of year dictates what and how training will occur. Wintertime may find you in the gym on a treadmill, stair climber or any other cardio-related torture device due to climatic conditions, depending on what part of the country you live in. In winter, electrolyte loss is not as much of a concern as in summer.

Hydration (consumption of fluids) is key for optimal performance. Your body's rate and individual reaction to the fluid loss from your workouts and breathing must be considered. To help you determine how much fluid you lose during a workout, weigh yourself before and after an hour of exercise. Drink enough fluids to replace at least 80% of that lost during exercise. Don't wait until you are thirsty. Although sports drinks can assist in the replacement of lost electrolytes during extended competitions, water can usually do the job as well or better.

The Grinder

(The drill ground, usually at a Fire Department training center, used for the PAT)

You don't have any control over the time of year, location or components of the physical ability test. It's important to consider what you *do* have control over.

An example of the need to do some thinking about the time and place of the exam would be as follows: the Biddle exam is given on August 15th at 1500 hrs on an asphalt drill ground with no wind and high humidity. The potential for optimal performance is already reduced. Couple this with minimal preloading of fluid and inadequate rest and food intake, and you are at a serious disadvantage.

72 hours prior to your appointment with the *grinder:*

In this time frame the amount of strenuous exercise should be limited to stretching, light running and resistance training, along with adequate rest. It is necessary to carbo-load (increase the intake of carbohydrates) to build up glycogen storage in your muscles to endure the test. The body needs to have both oxygen and glucose to adequately fuel and run the machine; without either one you can forget about any performance.

48 hours prior to competition:

Continue to increase your carbohydrate intake, as glycogen stores need to be maxed out for the big day. Light training with resistance or weights is advised, but no maximum repetitions or heavy lifting. Why, you may ask? What if you get injured just two days prior to your test date? The stiffness that normally accompanies a heavy workout may also hinder your times due to muscle fatigue and soreness.

While the muscle is healing from a heavy workout, its ability to "refuel" with carbohydrates is decreased. This means that no matter how many carbohydrates you eat, you simply can't get your muscle energy back up to normal for at least 48 hours after moderate to heavy exercise.

24 hours and counting:

Continue light and easy training. Organize your food and drink for the day of the test. Check directions to the Grinder. Have a high carbohydrate dinner, and get plenty of sleep.

Much like the military obtains intelligence on its adversary, you have targeted and done recon and intel on the desired department of your dreams. You, the serious candidate, have gone to the Fire Stations and spoken with the inhabitants (after offering gifts of ice cream, coffee, cookies, etc), and gotten the straight scoop from the rookie who last completed the physical ability test. Now loaded with mind comfort and confidence, the time has come to perform for a portion of the *four million dollar deal*. Four million dollars is what a Firefighter will earn in a thirty-year career.

Notes:

Sample Physical Ability Test

The following descriptions are examples of commonly used physical ability tests. These are printed here with permission from the Austin Texas Fire Department. Actual tests will vary from department to department.

The Fire Service Joint Labor Management Wellness/Fitness Initiative Candidate Physical Ability Test© consists of eight separate events. The CPAT is a sequence of events requiring the candidate to progress along a predetermined path from event to event in a continuous manner. This test was developed to allow fire departments a means for obtaining pools of trainable candidates who are physically able to perform essential job tasks at fire scenes.

Event 1: Stair Climb

Using a StepMill stair-climbing machine, this event is designed to simulate the critical task of climbing stairs in full protective clothing while carrying a high-rise pack (hose bundle) and firefighter equipment. This event challenges aerobic capacity, lower body muscular endurance and the ability to balance.

Participants wear a 12.5-pound weight on each shoulder to simulate the weight of a high-rise pack. Immediately following a 20-second warm-up period at a rate of 50 steps per minute, the timed part of the test starts as indicated by a proctor. There is no break in time between the warm-up period and the actual timing of the test. During the warm-up period, dismounting, grasping the rail, or holding the wall to establish balance and cadence is permitted. The timed part of the test lasts three (3) minutes at a stepping rate of 60 steps per minute.

Failure can occur by falling or dismounting three times during the warm-up period, or by falling or dismounting the StepMill after the timed CPAT begins. During the test, the participant is permitted to touch the wall or handrail for balance only momentarily; if that rule is violated more than twice during the test, failure will result.

Event 2: Hose Drag

This event is designed to simulate the critical tasks of dragging an uncharged hoseline from a fire apparatus to a structure and pulling an uncharged hoseline around obstacles while remaining stationary. This event challenges aerobic capacity, lower body muscular strength and endurance, upper back muscular strength and endurance, grip strength and endurance and anaerobic endurance.

A hoseline nozzle attached to 200 feet of hose is grasped and placed over the shoulder or across the chest up to eight feet. While walking or running, the participant drags the hose 75 feet to a pre-positioned drum, makes a 90° turn and continues an additional 25 feet. After stopping within the marked box, the candidate drops to at least one knee and pulls the hoseline until the 50-foot mark crosses the finish line.

During the hose drag, failure results if the participant does not go around the drum or goes outside of the marked path. During the hose pull, a warning is given if at least one knee is not kept in contact with the ground or if the knees go outside the marked boundary line; a second warning constitutes failure.

Event 3: Equipment Carry

This event uses two saws and a tool cabinet replicating a storage cabinet on a fire truck. It simulates the critical tasks of removing power tools from a fire apparatus, carrying them to the emergency scene and returning the equipment to the fire apparatus. This event challenges aerobic capacity, upper body muscular strength and endurance, lower body muscular endurance, grip endurance and balance.

The candidate must remove the two saws from the tool cabinet, one at a time and place them on the ground. Then he/she picks up both saws (one in each hand) and carries them

while walking 75 feet around a drum, then back to the starting point. Placing the saw(s) on the ground to adjust a grip is permitted. Upon return to the tool cabinet, the saws are placed on the ground, then picked up one at a time and replaced in the cabinet.

Dropping either saw on the ground during the carry will result in immediate failure. A warning will be given for running; a second warning constitutes a failure.

Event 4: Ladder Raise and Extension

This event, which uses two 24-foot aluminum extension ladders, is designed to simulate the placement of a ground ladder at a fire structure and extending it to the roof or window. This event challenges aerobic capacity, upper body muscular strength, lower body muscular strength, balance, grip strength and anaerobic endurance.

The participant must walk to the top rung of one ladder, lift the unhinged end from the ground and walk it up hand over hand until it is stationary against the wall. Then he/she immediately proceeds to the other pre-positioned ladder, stands with both feet within the marked box, extends the fly section hand over hand until it hits the stop, then lowers it back to the starting position.

Immediate failure will result if the ladder is allowed to fall to the ground, if control is not maintained in a hand-over-hand manner, or if the rope halyard slips in an uncontrolled manner. Missing any rung during the raise or allowing one's feet to extend outside of the boundary results in a warning; a second warning constitutes a failure.

Event 5: Forcible Entry

This event uses a mechanized device that measures cumulative force and a 10-pound sledgehammer. It simulates the critical tasks of using force to open a locked door or to breach a wall. This event challenges aerobic capacity, upper body muscular strength and endurance, lower body muscular strength and endurance, balance, grip strength and endurance and anaerobic endurance.

For this event, the candidate uses the sledgehammer to strike a measuring device in a target area until the buzzer activates. Feet must be kept outside the toe-box at all times.

Failure results if the participant does not maintain control of the sledgehammer and releases it from both hands while swinging. A warning is given for stepping inside the toe-box; a second warning constitutes a failure.

Event 6: Search

This event uses an enclosed search maze that has obstacles and narrowed spaces. It simulates the critical task of searching for a fire victim with limited visibility in an unpredictable area. This event challenges aerobic capacity, upper body muscular strength and endurance, agility, balance, anaerobic endurance and kinesthetic awareness.

For this event, the candidate crawls through a tunnel maze that is approximately three feet high, four feet wide, 64 feet in length and has two 90° turns and multiple obstacles. In addition, there are two locations where the dimensions of the tunnel are reduced. If at any point the participant chooses to end the event, he/she can call out or rap sharply on the wall or ceiling and will be assisted out of the maze although doing so will result in failure of the event. Failure also will occur if the candidate requests assistance that requires the opening of the escape hatch or opening of the entrance/exit covers.

Event 7: Rescue

This event uses a weighted mannequin equipped with a shoulder harness to simulate the critical task of removing a victim or injured partner from a fire scene. This event challenges aerobic capacity, upper and lower body muscular strength and endurance, grip strength and endurance and anaerobic endurance.

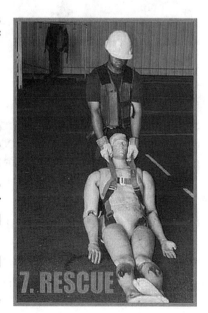

The participant grasps a 165-pound mannequin by the handle(s) on the shoulder(s) of the harness (either one or both handles are permitted), drags it 35 feet, makes a 180° turn around a pre-positioned drum and continues an additional 35 feet to the finish line. Grasping or resting on the drum is not permitted, but the mannequin may touch the drum. The candidate is permitted to drop and release the mannequin to adjust his/her grip. The entire mannequin must be dragged across the finish line.

Grasping or resting on the drum at any time results in a warning; a second warning constitutes a failure.

Event 8: Ceiling Breach and Pull

This event uses a mechanized device that measures overhead push and pull forces and a pike pole. The pike pole is a commonly used piece of equipment that consists of a six-foot long pole with a hook and point attached to one end. This event simulates the critical task of breaching and pulling down a ceiling to check for fire extension. It challenges aerobic capacity, upper and lower body muscular strength and endurance, grip strength and endurance and anaerobic endurance.

After removing the pike pole from the bracket, the participant places the tip of the pole on a 60-pound hinged door in the ceiling and pushes it three times while standing within the established boundary. Then, the pike pole is hooked to a 80-pound ceiling device and pulled five times. Each set consists of three pushes and five pulls; the set is repeated four times. A pause for grip adjustment is allowed.

Releasing one's grip or allowing the pike pole handle to slip does not result in a warning or constitute a failure. The candidate may re-establish his/her grip and resume the event. If a repetition is not successfully completed, the proctor calls out "MISS" and the apparatus must be pushed or pulled again to complete the repetition. This event and the total test time ends when the final pull stroke repetition is completed and the proctor calls "TIME."

A warning is given for dropping the pike pole to the ground or for feet straying outside the boundaries; a second warning of either violation constitutes a failure.

This is a pass/fail test based on a validated maximum total time of 10 minutes and 20 seconds.

In these events, the candidate wears a 50-pound vest to simulate the weight of self-contained breathing apparatus (SCBA) and firefighter protective clothing. An additional 25 pounds, using two 12.5-pound weights that simulate a high-rise pack (hose bundle), is added to the shoulders for the stair climb event.

Throughout all events, the participant must wear long pants, a hard hat with chin strap, work gloves and footwear with no open heel or toe. Watches and loose or restrictive jewelry are not permitted.

All props were designed to obtain the necessary information regarding physical ability. The tools and equipment were chosen to provide the highest level of consistency, safety and validity in measuring the candidate's physical abilities. A schematic drawing of the CPAT is included in this orientation material; however, the course layout may vary in order to conform to the fire department's test area. The events and distances between events are always the same.

The events are placed in a sequence that best simulates fire scene events while allowing an 85-foot walk between events. To ensure the highest level of safety and to prevent exhaustion, no running is allowed between events. This walk allows approximately 20 seconds to recover and regroup before each event.

To ensure scoring accuracy, two stopwatches are used to time the CPAT. One stopwatch is designated as the official test time stopwatch, the second is the backup stopwatch. If mechanical failure occurs, the time on the backup stopwatch is used. The stopwatches are set to the pass/fail time and countdown from 10 minutes and 20 seconds. If time elapses prior to the completion of the test, the test is concluded and the participant fails the test.

Being physically fit is extremely important to being able to perform your job.

Test Forms

Prior to taking the CPAT, each candidate must present valid identification, sign a number of forms, complete a waiver and release form and a sign-in form. Candidates are provided an opportunity to review a video detailing the CPAT and the failure points. It is the candidate's responsibility to ask questions if any part of the test events or procedures are not understood. At the conclusion of the CPAT, the candidate must sign the CPAT Evaluation Form and complete and sign the Rehabilitation Form. Failure to complete and sign any of these forms results in failure of the CPAT.

Background Investigations

Fire departments traditionally spend thousands of dollars to advertise, recruit and hire firefighters. The departments sift through applicants using written examinations, physical ability tests and comprehensive oral interviews, but only do a cursory check on their backgrounds. They eventually produce a list of top candidates. It is now up to the organization to ferret out those candidates who were less than truthful on their application or during their interview.

Background investigations are an important component of the hiring process. They are completed by most fire departments across the country. Historically, fire departments have not placed as much emphasis on a thorough background check as their counterparts on the police department. A criminal check with the local police agency and a DMV check was the extent of what we used to look at.

The local police departments often complete today's background checks. Many fire departments hold their firefighter candidates to the same high standards expected of a police officer. These standards include criminal history, drug usage, credit history, employment record, encounters with the law and a candidate's overall persona.

The reasoning is that if a person has demonstrated an inability to manage his or her personal finances, is unable to get along with co-workers, or has simply made poor life decisions, these will be magnified as their responsibilities increase. If, on the other hand, a candidate has demonstrated a strong history of being able to manage his or her personal and professional life, there is no reason to expect that he or she would not continue to do so after being hired by the agency.

Gordon Graham, an attorney and well-known expert on issues pertaining to police and fire departments, believes that "The best predictor of future behavior is past behavior." He feels that if a candidate has had problems in the past, he or she will have problems in the future. His advice to police and fire chiefs across the country is, "Why take the chance and incur the liability, especially when you have so many candidates to choose from." A thorough background check can help an agency reduce its future incidents of personnel problems and minimize the risk of negative publicity for the agency. Patterns of past performance issues and problems with co-workers are a strong indicator of future behavior and should not be overlooked.

A thorough background investigation is important because of the role of the fire department in the community. The firefighter candidate will eventually hold a position of authority and responsibility. Firefighters are welcomed into people's homes and businesses without fear for their personal safety or their prized possessions. If the candidate is of questionable ethical or moral character, he or she may ultimately become a liability for the hiring agency. This could erode public trust and compromise the department.

The U.S. Chamber of Commerce estimates that dishonesty by employees costs a business 1-2% of its gross sales. Surveys reveal that 33% of employees admit to stealing product or money from their jobs in the last three years. It is estimated that 30% of businesses fail because of employee theft. Statistics also reveal that roughly 40% of applicants have false information on their applications.

Negligent hiring litigation is on the rise. Employers lose 72% of all negligent hiring suits, with the average award to the plaintiff exceeding one million dollars. Most of these are due to the employer failing to take the proper steps to avoid hiring an unfit employee. Courts have ruled that an employer has a general duty to check criminal records for employees who will interface with the public.

Once a candidate has been selected to move on in the hiring process, he or she is assigned a background investigator. Before meeting the background investigator, the candidate is given a background packet. These vary slightly from agency to agency and are often 25-30 pages long. A candidate is usually given 14-21 days to complete the packet prior to the first meeting with the investigator. Candidates are advised to photocopy the packet and fill out the copy in pencil. Once the rough draft is complete, the original is completed in pen or, even better, typed.

Neatness is a characteristic that is important to a background investigator. If he or she is unable to decipher an applicant's chicken scratch, it makes a poor first impression. A typed background packet, on the other hand, gives the impression of being thorough and complete.

The background packet will seek information relating to all jobs held (including names of supervisors and dates employed), military record including DD214, sealed high school and college transcripts and a thorough questionnaire regarding drug and criminal history. Applicants will be expected to complete a section that outlines any and all encounters with illegal drugs, including persons involved, dates and times, as well as the number of times he or she has experimented with each substance.

Any omission of information is considered to be covering up and will be viewed as deceitful, which is grounds for automatic disqualification. If a candidate legitimately forgets information, it can certainly cost him or her a job. To avoid making these costly mistakes, a candidate should keep a log of information that would be helpful to a background investigator, including names and addresses of landlords, employers, friends and co-workers. Any blanks left in the packet raises the question of whether the applicant is attempting to cover something up.

Once the applicant has completed the background packet, he or she will be scheduled to meet with the assigned investigator. The investigator may be a firefighter on the department, a police officer for the city or county, or a private contractor. Whoever it is, the applicant's future employment relies on successfully completing the process.

The investigator will take several photos of the candidate that will be shown to friends, neighbors and co-workers during the investigation. The applicant will be asked for a list of friends and close associates, including their names, addresses and phone numbers. The prospective firefighter must sign a stack of release waivers that will be used by the investigator for each person contacted.

The investigator will review the background packet with the applicant, seeking to identify any discrepancies and delve deeper into them. This is the applicant's opportunity to explain his or her side of what transpired. It is akin to going to confession. After this stage, anything uncovered by the investigator that was not previously disclosed is considered to be intentionally "forgotten" and could be used as a foundation for dismissal from the hiring process. Once an investigator gets a feel about a candidate from the interview, he or she will begin some cursory checks of driving and criminal records, as well as a credit check.

Driving records are important, since having a current driver's license is required for most firefighter positions. A candidate who has a history of speeding or ignoring traffic laws may be disqualified since we operate emergency vehicles. Driving lights and siren through the city is a huge liability for the agency. Imagine if a firefighter was driving lights and siren at an excessive rate of speed and plowed into a bus bench full of school children. The subsequent investigation revealed that the firefighter had a series of speeding and moving violations. The agency would probably lose any lawsuit. Even if it didn't, it would certainly be a black eye for the department.

Numerous parking tickets make a statement of how a candidate reacts to authority. If a candidate has a series of infractions (paid or not), it could indicate that he or she feels that it is unnecessary to abide by society's rules.

I was asked in a seminar recently if a candidate would be held liable for the parking tickets he was given while driving the company delivery vehicle. He tried to reason that they weren't his fault because, as a delivery driver, his boss gave him permission to park in the red zone. I asked him who gave his boss the authority to tell him it was OK to ignore the law. He continued to tell me that since his boss told him it was OK and the company paid for the tickets, he felt he was off the hook. I told him that even if he were off the hook for the parking tickets, he would probably fail the background because he has a pattern of exercising poor judgment.

Another candidate asked if it would look badly if he was always the one to bail his friends out of jail. He rationalized that it showed he was a loyal and dedicated friend. He stated that he knew the fire service valued strong friendships and looking out for each other. I assured him that he was correct on both counts. We do value strong friendships and we certainly take care of each other. I would question why he is associating with people who are constantly being thrown in jail. I reminded him of the old saying, "Birds of a feather flock together." In other words, if your friends and associates are guilty, then so too are you. Whether this is the case or not is irrelevant; you define yourself by the company you keep.

Obviously, criminal records are important to the hiring agency. Firefighters routinely find themselves unsupervised in people's homes and businesses. Imagine for a moment the headlines in the local newspaper: "Firefighter arrested for stealing from elderly lady's bedroom while she was having a heart attack." Of course, this would be picked up by the national media and would be a black eye for all firefighters.

Credit history is also important, as it too makes a statement of how an individual handles responsibilities. If a person is not able to live within his or her means, this person is a potential liability to the agency. A blemished credit history may indicate an inability to handle responsibility.

Bankruptcy is a big red flag to an agency. Simply because a credit card company considers an individual untouchable and relieved of financial responsibility once he or she declares bankruptcy, fire departments do not view this in the same way. In reality, although an individual has declared financial bankruptcy, he or she is morally obligated to repay the money that was borrowed. In the eyes of the law the obligation has been "forgotten," but somebody is still out money. Is it an automatic disqualification? No, not if there has been progress made toward repaying the debt after bankruptcy was declared. According to a former background investigator for LAPD, "If an individual is making an honest effort to repay the money, we can look past a bankruptcy. We cannot overlook someone who does not attempt to right known wrongs."

Candidates often wonder if they should report things that occurred when they were younger. They feel that if a record was sealed, they are not accountable for anything until they reached 18 years of age. Nothing is further from the truth. Remember the forms you signed when you sat down with the background investigator? These give permission to look into every aspect of your life. There is no such thing as a sealed record to a background investigator. Even if there were, whatever a candidate did to get a police record sealed would be cause for alarm and would raise the issue of liability for the agency. For the record, there is no such thing as a sealed file, regardless of what your attorney tells you.

Many people believe that they can give the background investigator only the names of their responsible friends, the ones who will say positive things about them. They will make sure to brief their friends on what to say and what not to say. In effect, they will coach them on how to answer the questions. Certainly, the investigator will interview the people listed by the candidate, but

they will also ask the individual for the name of five friends. They will interview the five new people and when completed, will ask for five more friends and so on. It doesn't take long for a trained investigator to get to someone who has not been coached.

The investigator will knock on the door of your neighbors and show them a Polaroid picture (the same one taken on the day of your initial background interview). If your neighbor tells the investigator that it looks like you, but the nose ring and bandana that you always wear are missing, the cat is out of the bag. In other words, the investigator has learned a lot about you. Will this disqualify you? Probably not, but it now gives the investigator cause to look deeper into your profile.

This scenario is the number one reason that when I speak to a group of fire science students, I encourage them to look the part. You don't see many firefighters with nose rings and bandanas. The students constantly assure me that when they start testing, they will shave off the goatee and get a haircut. It is important to note that we are not looking to hire the person who can do a complete makeover in 30 days or less. If you changed that quickly to get the job, it stands to reason that you will change back after you get it. We are looking to hire people who authentically live their lives in a positive fashion.

Are there automatic disqualifications for the background process? Yes and no. What does this mean? It depends on the agency and on the feelings of the fire chief. Some fire chiefs don't care what you have done (within reason), but will automatically disqualify a candidate who is not completely honest during the process, while others have certain actions that are immediate cause for dismissal.

Some common causes for automatic disqualification include the following: any injectible drug use (i.e. any controlled substance or steroid put into the body via a needle); any selling, intent to sell or transporting of narcotics; use of hallucinogens such as LSD and acid; multiple uses of marijuana that is considered more than experimental; any type of assault or domestic battery; stealing and arson.

Of course, these are generic, but most agencies will have a policy dealing with any of the above cases. For some it may be an automatic disqualification, while other agencies may be more lenient and receptive to a reasonable explanation.

If a candidate has a blemish on his or her record that is not considered an automatic disqualification, the investigator will look further into the background. The intent is to determine if the infraction is a one-time incident or a pattern of poor choices. Oftentimes a driving under the influence arrest was the proverbial accident waiting to happen. In other words, a candidate tells the investigator that after the annual company picnic, he or she had too much to drink. The designated driver was nowhere to be found and the candidate had to get home to feed his or her cats. It was a matter of life and death. The candidate got behind the wheel and drove when he or she shouldn't have. As luck would have it, the candidate rear-ended a police car and was arrested for driving under the influence. It was just an isolated incident that could have happened to anybody, right?

This would naturally trigger the investigator to look further into the candidate's alcohol consumption. In fact, one of the questions is, "How often do you drink?" Nobody wants to look like an alcoholic, so they grossly underestimate the number of times alcohol is consumed each week. This is easily uncovered by interviewing your friends, who vouch for the fact that you are able to hold your liquor.

When it is revealed that you play softball in a beer league every Tuesday night with the guys from the shop, the investigator will easily identify that you drink every Tuesday night. Of course, the next question will be, "How does the candidate get to and from the game?" Your helpful friend raves about the pickup truck that you completely restored and drive to each and every game. The connection is now made complete that after drinking during the weekly softball game, the candidate hops into his restored pickup and drives home. Now, the driving under the influence conviction is no longer an isolated event, but rather part of a pattern of poor choices that finally caught up with a careless individual.

If, on the other hand, it does appear to be an isolated event, the investigator will want to know what you have learned from the event. A candidate who was arrested for driving under the influence four years ago, has since quit drinking and is now a designated driver on the major holidays and a spokesperson for Mothers Against Drunk Driving (MADD), will certainly be considered above the previous candidate. In this scenario, it's not the mistake that draws the attention, it's the recovery.

A person who has smoked marijuana is usually not eliminated unless it was done in recent history. Some departments will eliminate a candidate if it was done after the candidate decided he or she wanted to become a firefighter. Again, an example of poor decision making. In today's day and age, it is understood that most people will at least try marijuana. In fact, a recent news study revealed that 66% of high school seniors have at least tried it. Unfortunately, it seems to be on the rise. If smoking marijuana were an automatic disqualification, the fire and police agencies across the country would not be able to hire most new employees. The applicant pool would simply be too small.

The background investigation is the time to be accountable for all of your life's actions. Most people have some past indiscretions that, if given the choice, they would change. That is what we call life experience. If the individual is honest and forthcoming with information and has not made any life-altering decisions, as mentioned above, he or she should have no problem passing a comprehensive background check. It is important to note that if a candidate believes he or she may have difficulty with a background investigation, he or she probably will.

My advice is to be honest and forthright with information. Everyone makes mistakes. If a candidate is caught in a lie, he or she is automatically eliminated from the process, even if the issue was not a big infraction. The fact that the candidate lied says a lot about his or her character.

Once a candidate fails a background investigation, the next agency has a right to know about it. In other words, when a candidate goes through a background investigation for a different agency, they have a right to know why you failed. If a candidate failed a background for lying, chances are they will not make it through the next process.

Background Investigation Questionnaire

Read and answer the following questions **carefully** and **honestly**. Answers are subject to verification by a Polygraph Examination.

Have you _ever_ committed any of the following acts _during your life_ whether it came to the attention of authorities or not?

		YES	NO
1.	Spousal abuse (including common-law)	❏	❏
2.	Any violent assault upon another	❏	❏
3.	Forgery	❏	❏
4.	Homicide	❏	❏
5.	Robbery (theft from another person utilizing a weapon of force)	❏	❏
6.	Burglary	❏	❏
7.	Kidnapping	❏	❏
8.	Arson (intentionally set fire)	❏	❏
9.	Extortion (blackmail)	❏	❏
10.	Embezzlement (theft of money)	❏	❏
11.	Rape (sexual intercourse by force or against the wishes of another)	❏	❏
12.	Child Abuse	❏	❏
13.	Child Molestation (any sex act with a child)	❏	❏
14.	Prostitution (sexual acts for money or other considerations)	❏	❏
15.	Soliciting Prostitution (asking for sex in return for money or visa versa)	❏	❏
16.	Theft (including shoplifting)	❏	❏
17.	Convicted of a DUI or charge reduced to reckless driving	❏	❏
18.	Have you in the past or do you now regularly associate with persons whom you know to have engaged in and/or been arrested for unlawful possession or use of any illegal substance?	❏	❏
19.	Have you ever been arrested for an illegal sex act?	❏	❏
20.	Ever charged with a crime not mentioned above?	❏	❏

Explain all "yes" answers to the above questions on the following page.

Explain all "Yes" answers to the above questions in the spaces below. List the question # and then describe the incident. Be thorough with your explanation; use dates where appropriate.

_____ _____

_____ _____

_____ _____

_____ _____

_____ _____

_____ _____

_____ _____

_____ _____

_____ _____

_____ _____

_____ _____

_____ _____

Employment

		YES	NO
21.	Have you ever been terminated or asked to resign from employment?	❏	❏
22.	Have you ever taken *any* property from an employer?	❏	❏
23.	Have you in any way falsified your employment application or made any misleading statements?	❏	❏
24.	Have you ever made any false or misleading statements or omissions to any employer or potential employer?	❏	❏
25.	Have you committed any dishonest act in order to obtain this or any position (i.e. cheating on written exam, having another person take any exam, etc.)..	❏	❏

Explain "Yes" answers in detail below. List the question # and then thoroughly explain.

_____ _____

_____ _____

_____ _____

_____ _____

_____ _____

_____ _____

_____ _____

_____ _____

Financial

	YES	NO
26. Do you feel that you now have a good credit rating (If yes, no explanation necessary)?	❏	❏
27. Do you have any bills that are now past due?	❏	❏
28. Have you filed for bankruptcy within the past two years?	❏	❏
29. Have you ever failed to file an income tax return?	❏	❏
30. Have you had a bill turned over to collections within the past two years?	❏	❏

Explain all "Yes" answers to the above questions in the spaces below. List the question # and then describe the incident. Be thorough with your explanation; use dates where appropriate.

_____ _____

_____ _____

_____ _____

_____ _____

_____ _____

_____ _____

_____ _____

_____ _____

_____ _____

General

31. Did you answer all of the questions truthfully that were put to you during the oral interview? .... ❑   ❑

32. Are you presently driving without auto insurance or the DMV required Financial responsibility bond? .... ❑   ❑

33. Have you ever filed a fraudulent insurance claim? .... ❑   ❑

34. How many traffic citations have you received within the past five years?... ❑   ❑

35. Have you ever taken a Polygraph Examination?.................... ❑   ❑

When _____ Where _____ Result _____

36. Have you ever been present when someone else committed a criminal act?.. ❑   ❑

37. Have you ever purchased or sold any property that you believe might have been stolen?... ❑   ❑

Explain all "Yes" answers to the above questions in the spaces below. List the question # and then describe the incident. Be thorough with your explanation; use dates where appropriate.

_____   _____

_____   _____

_____   _____

_____   _____

_____   _____

_____   _____

_____   _____

Drugs/Narcotics

	YES	NO
38. Have you used marijuana?	❏	❏

Last time: Month/Year _____ First time: Month/Year _____

	YES	NO
39. Have you ever misused a prescription drug?	❏	❏
40. Within the past year, have you been in the presence of anyone using illegal drugs?	❏	❏
41. Have you ever purchased, sold, or supplied any illegal narcotic, steroid, marijuana, pill or drug?	❏	❏
42. Have you ever been the middleman, go between, or "done a favor for a friend" by becoming involved in a drug transaction?	❏	❏

43. Have you *ever* used or experimented with:

	YES	NO
Cocaine	❏	❏
Heroin	❏	❏
LSD (acid)	❏	❏
PCP (angel dust)	❏	❏
Mushroom, mescaline or any hallucinogen	❏	❏
Hashish	❏	❏
Crank, Methamphetamine	❏	❏
Speed or Crystal Meth	❏	❏
Uppers	❏	❏
Downers or Barbiturates	❏	❏
Steroids	❏	❏
Ecstasy	❏	❏
Any type of "designer" drugs	❏	❏
Any other drug besides Marijuana	❏	❏

If you answered "Yes" to any of the above drug usage questions, complete the section below:

Drug_____ First Time_____ Last Time_____

Drug_____ First Time_____ Last Time_____

Drug_____ First Time_____ Last Time_____

Polygraph Exams

The intent of this chapter is not to advise candidates how to "beat" a polygraph exam, but rather to educate them on the process. Much of the research for this chapter was conducted by interviewing candidates who have been through the process and via my own personal experience with a pre-employment polygraph examination.

The name "polygraph" refers to the manner in which physiological activities are simultaneously recorded. The term literally means "many writings."

The Employee Polygraph Protection Act of 1998 (EPPA) prohibits most private employers from administering prehire polygraph interviews. Of note, however, is the fact that this does not apply to public employers such as police and fire departments or governmental institutions.

Polygraph examiners use a series of different instruments placed strategically on the subject's body. Convoluted rubber tubes that are placed over the examinee's chest and abdominal area will record respiratory activity. Two small plates attached to the fingers will record sweat gland activity and a blood pressure cuff will record cardiovascular activity.

Typically, polygraph examiners will administer a "pre-test." During this period, the examiner will complete required paperwork and discuss the questions that will be covered during the exam proper. It is not uncommon for the examiner to ask the subject to intentionally lie about his or her age. The examiner shows the physical results to the subject to lend credibility to the test. Following the initial interview, the examiner will question the subject on anything that gives an unusual reading. The subject will have an opportunity to explain any unusual findings.

Proponents believe polygraph examinations are extremely accurate, while opponents argue that there is minimal science associated with the process. According to the Journal of Applied Psychology (1997), the polygraph examination has a 61% accuracy rate. According to Jerry Smith, former CIA general counsel, "The polygraph is not perfect. Honest people have failed, while dishonest people have passed. The polygraph is intrusive and may be abused. If it is misused it can ruin the careers of honest people."

In an American Civil Liberties Union briefing paper, the article explains that despite the claims of lie detector examiners, there is no reliable machine that can detect lies with any degree of accuracy. The "lie detector" does not measure truth telling; it measures changes in blood pressure, heart rate, breathing rate and perspiration. A wide range of emotions, such as anger, sadness, embarrassment and fear, can trigger these physiological changes. In addition, a variety of medical conditions, such as colds, headaches and neurological and muscular problems, can distort the results. Indeed, as an American Medical Association expert testified during public hearings before Congress, "The lie detector cannot detect much better than a coin toss."

Proponents of polygraph screening believe that applicant's prior knowledge of the agency's policy to administer a polygraph results in a higher caliber of applicants. In other words, the ones with questionable backgrounds do not even apply. In addition, the applicants who do apply are generally more honest, knowing they will be put to task. Of course, this belief cannot be verified.

The American Polygraph Association (APA) Research Center at Michigan State University conducted a study of police departments to determine the extent of polygraph use for pre-employment screening for police officers. The survey included roughly 700 of the nation's largest police departments, excluding federal agencies. The results revealed that 62% of police departments administer pre-employment polygraph examinations, while 31% did not and 7% had discontinued the use because of legislation that had been put into place within their jurisdiction.

Of the applicants tested, roughly 25% were disqualified. Although it is difficult to determine exactly why the applicants were disqualified, the overwhelming majority were disqualified for some form of serious undetected crime. Of the agencies surveyed, the polygraph screening revealed that 9% were involved in unsolved homicide, 34% had some involvement with forcible rape and 38% had participated in armed robberies.

According to the APA and the EPPA, no examiner shall delve into the following: religious beliefs; opinions or beliefs regarding racial matters; political beliefs or affiliations; lawful activities or affiliations with labor unions or labor organizations; sexual preferences or activities. Similar questions are presumably asked in fire department pre-examinations.

In law enforcement pre-employment examinations, the questions focus on such job-related inquiries as the theft of money or merchandise from previous employers, falsification of information on the job application, the use of illegal drugs during working hours and criminal activities. Similar questions are presumably asked in fire department pre-employment examinations.

The results of the polygraph examination can only be released to authorized persons. These are generally considered to be the examinee and the person, firm, corporation or governmental agency which requested the examination.

If a polygraph examinee believes an error has occurred in the process, he or she has several options. He or she should first request in writing a second examination and retain an independent examiner. In the fire department testing arena, this would certainly come out of the applicant's pocket with no guarantee the agency would accept the results. The applicant may also choose to file a complaint with the state licensing board for polygraph examiners and the Department of Labor. Lastly, he or she may file a request for assistance from the American Polygraph Association.

Polygraphs are not an exact science. At best they can give the examiner a strong indication the examinee may not be telling the truth. At worst they can give a false reading, which may ultimately result in declaring the applicant to be telling a lie. Whichever the case, applicants need to educate themselves on the process, as they are becoming more popular. An Internet search under "polygraph" examinations should yield more information for those who are interested.

Notes:

Learn from other's mistakes; you'll never live long enough to make them all yourself.

Personal Attributes

Being a strong communicator is an important part of the job of a firefighter.

Public Speaking

Firefighters are always in the public eye. It is a vital part of the job. In the hiring process, the interview is usually given the highest number of points. In other words, fire departments across the country are looking for people who are comfortable with public speaking.

Studies have shown that people's greatest fear is speaking in front of a group. Steve Lewis, the Chief of Newport Beach Fire Department in Southern California, says, "If you are a good public speaker, I cannot guarantee you a job in the fire service. If, on the other hand, you are not a good public speaker, I can guarantee you will never get a job in the fire department." While this may be a bit extreme, Chief Lewis advises candidates to take public speaking courses at their local community college. In addition, he is an advocate of Toastmasters. Wherever they get the experience, candidates who become good public speakers will always fare better than those who struggle with it.

Some people have the gift of gab. These candidates usually score well in the process since they are used to speaking in front of a crowd. Imagine for a moment the person who is the life of the party. He or she is used to having to communicate in front of an audience and are a natural at performing on stage.

These people don't necessarily make better firefighters; they just stand out in the hiring process. And since it is important to learn the rules of the road, a wise candidate will learn how to speak in front of a group.

It is important to have firefighters who can think on their feet. We deal with the public literally dozens of times each day. These encounters range from a brief encounter in the grocery store, to speaking engagements for the local grade schools, to being assigned to put on a presentation before the city council or board of fire commissioners.

One of the most important aspects is speaking to a family member about the condition of his or her loved one who may be ill. It is imperative that we are able to effectively communicate to establish patient confidence. Imagine how you would feel if the doctor who was treating your mother was socially illiterate. Naturally, you would question his or her ability.

How do you become a good public speaker, especially if it is difficult for you? The best way to conquer any fear is to meet it head on. In other words, put yourself in a position that forces you to perform in public. Maybe it is a matter of joining a committee at work that forces you to do presentations. Another way is to volunteer for an assignment that involves training your peers. If you can perform in front of your peers, you can perform in front of anyone.

For me the answer was teaching first aid and CPR courses. It forced me to get in front of a group of people. Most importantly, I became very proficient in my EMT skills. Imagine having to answer questions from 10 students in a CPR class. You quickly learn to formulate your thoughts and present them in an organized manner. After some time teaching in front of a group, you will find that your fears of public speaking have subsided. Now instead of being nervous when having to answer questions from 10 people in a CPR class, you are only fielding questions from three people.

In order to improve your public speaking skills, it is important to understand the characteristics of a good public speaker. While some of these examples directly relate to the interview process, others relate to public speaking in general.

Body Movements

- Researchers say that words account for only 35% of what we communicate; the rest is largely accomplished through body language.
- Practice gesturing, but not specific gestures – it looks artificial.
- Don't fidget (e.g. playing with a pencil, etc.)
- Move when there's a reason to move.
- Stand still when there's not a reason to move.
- As you watch television, observe how speakers use gestures as they talk.
- Don't make up gestures that you don't feel comfortable with.

Six Traditional Speech Gestures:

1. Giving and taking – hand out with palm up.
2. Raising a fist – shows strong feelings; be careful with it.
3. Pointing – indicates position; calls attention to something; makes an accusation.
4. Rejecting – sweeping gesture with your hand, palm downward.
5. Dividing – palm in a vertical position, moving it side to side. Conveys parting ideas and allows new ones to form.
6. Warning – place your hand straight out like a stop sign, with palm out, heel of the hand down. Can also calm an audience or prepare listeners to accept another idea.

When using gestures, make them strong and accurate. The audience should know what they are being used for.

- Establish eye contact with the audience.
- Eye contact with individuals should not be for more than one sentence.

If using gestures is not part of your normal makeup, don't panic. Your speech can still be successful.

- Clasp your hands as a steeple or spider on a mirror at waist level.

- Remember, if you don't have a lot of credibility, then your delivery becomes critical.
- You are the most important visual.
- If you're enthusiastic, your audience will be too.
- Simply smiling at an audience can create instant rapport.
- You don't have to smile the whole time.

Learn to use your face. Raise your eyebrows in disbelief; frown if your speech calls for disagreement.

Posture DO'S and DON'TS

- Don't lean on the podium.
- Don't stand with your hands on your hips.
- Don't sway back and forth.
- Don't stand with your arms folded across your chest.
- Don't stand with your arms behind your back.
- Don't stand in the fig leaf position.
- Don't bury your hands in your pockets.
- Don't play with the coins in your pockets.
- Don't keep adjusting your glasses or taking them off and on.
- Don't play with your jewelry.
- Don't invade someone's space.
- Don't physically touch the audience.
- When practicing, use large exaggerated gestures to get the feel.
- Do stand up straight with your feet slightly apart and your arms at the ready.
- Do lean slightly toward the audience.

Eye Contact

- Don't speak unless you have eye contact with the audience.
- Don't keep directing all your attention towards the one friendly face.

- Don't look out the window.
- Don't look at one spot.
- Don't forget to look at the back rows.
- Don't let notes ruin your eye contact.
- Don't look over the heads of the audience.
- Do establish eye contact at the end of a thought.
- Do look at individuals.
- The rule of thumb for eye contact is one to three seconds per person.

Impromptu Speaking

Some call it impromptu, speaking off the cuff, or extemporaneous speaking. Most people call it Hell.

You already know how to do it. It's called normal conversation.

Remember the following points:

Be prepared. Think that there is always the chance they'll call on you.

Take a few moments to get your ideas focused.

Stall creatively.

Pause thoughtfully – This technique actually increases your credibility. The audience thinks you're mulling it over before you speak.

Repeat the question – This buys you some time while you're organizing things in your head.

Open with a broad generalization to buy a few extra seconds to organize your thoughts.

Some say your mind goes blank. Not true. What usually happens is you're overwhelmed with thoughts.

Don't apologize for being caught off guard; they already know you're winging it and they'll cut you some slack.

Decide on the conclusion you want to present and organize your short talk around that.

Organize around a standard pattern:

1. Past, present, future
2. Problem, solution
3. Cause and effect
4. Pick a topic, add a few subtopics

Support your views with specifics.

Tie your comments into other speeches that have been given.

If the audience doesn't know that you're speaking off the cuff, tell them.

Tell a personal anecdote to make the main point.

Create a clear analogy to make your point.

If you're thoroughly stuck, reply with: "I don't know; I'll do some research on the topic and get back to you with what I find."

Again, don't apologize! If you do, you are basically telling your listeners to disregard, or at least discount, what you are about to say.

Audiences tend to be very forgiving of an impromptu speaker.

Your listeners expect a few hesitations, pauses, repetitions, rephrasing or silences.

Perfection in an impromptu speech is impossible. Don't try to achieve it.

Don't ramble on.

Don't get off your subject: "Oh, that reminds me of something else."

Don't act surprised.

Keep it simple.

When you're done – Stop and sit down!

Mark Twain once said, "*It usually takes me more than three weeks to prepare a good impromptu speech.*" (Joke)

Mechanical ability is an important quality for firefighters to possess.

Mechanical Ability

I was approached by one of the training captains on our department to spread the word to potential recruits to pay attention to learning one of the most basic components of our job, mechanical skills. He expressed that the average entry-level recruit has plenty of degrees and many classes and many are even certified paramedics. Unfortunately, most are lacking mechanical ability. He was quick to stress that the job is a hands-on mechanical job.

Since computers control all of the cars that are being built today, the days of growing up tinkering on your car in your parent's garage are a thing of the past. As a result, many of the current generation have spent their time playing video games. When it comes to mechanical ability, they are severely lacking. In a recent academy our drillmaster asked how many of the recruits

Being familiar with basic power tools is a great asset.

had never changed the oil in their own car. Shockingly, 30% of the recruits had never done a basic oil and filter change on a vehicle.

Since the general public dials 911 when they can't solve their problem, it is imperative that firefighters are good problem solvers. Although many of the calls for service are EMS-related and require little mechanical ability, there are still numerous times throughout the day that a firefighter must think outside of the box.

Firefighters respond to traffic accidents that require the car to be dismantled around a human body without causing further harm to the victim. We have state-of-the-art hydraulic tools and jacks which work wonders when used in the right manner. In the hands of an untrained professional they are dangerous. Complex extrications require firefighters to be resourceful and use their heads to solve problems. If a firefighter does not have a strong mechanical background, he or she will have little to offer.

Firefighters respond to calls for electrical shorts. A firefighter should understand the basics of how a house is wired. They should at least know how and when to shut down the electricity to a house or commercial building.

Natural gas leaks in a structure kill or injure hundreds of people each year. The fire department is the emergency response agency for the gas department in the community. Why would the fire department respond to a gas leak? Fire stations are situated throughout the community and are ready to respond at a moment's notice. They have the proper training and equipment to mitigate the problem. Even if they can't solve the problem, firefighters have the ability to identify the problem and make sure the area is safe before the gas department arrives, which is often an hour (or several hours) later. A firefighter candidate should know how and when to shut down the natural gas or propane that powers a house, apartment building, mobile home or commercial building.

A firefighter is expected to be able to tear down and rebuild basic gasoline motors that power the smoke ejectors and various saws used by the department. A basic understanding of gas-powered engines is helpful.

Here are a few suggestions of what you can do to enhance your mechanical ability:

1. Mow your lawn. Understand the basics of how a lawnmower works. The motors on a lawnmower are similar to a smoke ejector. The same theory applies to a chain or rotary saw. If you understand the basic concept of how the motor works, you will be able to tear it apart and rebuild it.

2. Change the oil in your car. Become familiar with the operation of basic hand tools. The more you use them, the more you will be able to adapt to changing situations on the fire ground or during a complex extrication. It is critical to be able to think outside of the box.

3. Buy a skill saw and build a doghouse. By doing so you will understand the basic operation of the saw, as well as have to plan the doghouse layout in your head.

4. Get a job in the construction industry while you are testing for the fire department. You will learn more that will apply to being a solid firefighter by doing drywall, setting tile, carpentry or plumbing than you would as a bank teller or computer technician. Even if the job involves construction demolition or hanging drywall, you will be exposed to how buildings are put together. The more you understand these basic facts, the safer you will be while cutting a roof at two in the morning in heavy smoke.

Many firefighter candidates believe that working as an EMT driving an ambulance will help them become a firefighter. In theory they are correct, but the reality of the situation is that an EMT does relatively little patient care. If it is an ALS (Advanced Life Support) run, the paramedics work on the patient. When it comes time for the patient to be loaded into the ambulance, the medics step back and the EMT's move in and figure the logistics of putting the patient on the gurney. Once the patient is in the back of the ambulance, the medics

*Firefighters must be comfortable working while
wearing self-contained breathing apparatus.*

resume patient care. With the exception of taking an occasional blood pressure, the EMT's are relegated to tearing tape for an IV or splinting fractures.

Is the EMT benefiting from the "field" experience? Yes, absolutely. If it were up to me, I would rather see the candidate working in the construction field learning about tools that he or she will ultimately use on the fire department. A part-time job on an ambulance working weekends and holidays as vacation relief will take care of the basic EMT skill maintenance.

The more experience a candidate has with basic mechanical concepts, the better he or she will perform in the academy and as a firefighter. If a candidate does not possess basic skills, he or she can learn them. It is true that some people have a basic knack for how things work, but it is also a learned skill. Certainly those who tinkered in the garage with their fathers or grandfathers have an advantage over those of us who had to learn it on our own.

In our academy, recruits are expected to put on a blacked-out mask and crawl out of a building. In the middle of the evolution, the candidate will

encounter a smoke ejector, a chain or a rotary saw. The recruits are instructed to start one of the three pieces of equipment while wearing the blacked-out mask, full turnouts and a breathing apparatus. The entire evolution is done while wearing heavy structure firefighting gloves. The recruits have three minutes to complete the evolution.

According to the training captain, the recruits who possess mechanical ability prior to entering the academy excel, while those who don't often struggle.

There are books geared for aspiring firefighters that discuss the basic principles of how things work. They can be purchased at any of your larger bookstores. I would highly recommend that if this is an area of weakness, you look to minimize your weakness. The more you understand, the easier it will be to learn your chosen career.

Learning mechanical skills from a textbook is a near impossibility. There is no substitute for hands-on training. Every time you get an opportunity to see something repaired, you will improve your basic understanding of how things work. After seeing a variety of things being repaired you will begin to see the commonalities of how things are put together. It is impractical to expect that you will ever learn enough to be a mechanic; however, you can learn enough to make you a competent firefighter.

If you already possess mechanical ability, you should emphasize it during the testing process, in particular during the oral interview. The oral board is generally made up of "older" firefighters, usually captains and above. We truly appreciate a new firefighter with a strong mechanical and construction background; it is up to you to bring it to our attention.

I can assure you that a candidate who has a strong mechanical background can make up for a lack of firefighting or EMS experience. We can teach you how to be a good EMT or paramedic. We cannot teach mechanical aptitude or common sense. That has to come from within.

The more a candidate can improve his or her mechanical ability, the easier the transition will be from civilian to firefighter. A basic fire academy at the junior college level will give a candidate a basic understanding of what is expected from a firefighter in terms of mechanical knowledge.

*When was the last time you
did something for someone else?*

Community Service

One of the important aspects of being a firefighter is community involvement. This comes in many forms. To some it may be as a little league or soccer coach, while others are involved in the Boys and Girls Club. Some may be active at church. Whatever you choose to do, it is important to be active in the community.

Many candidates tell me that they have so much going on, they do not have time to get involved in their neighborhood. Although they would like to make a difference in the community, there are just not enough hours in the day. What is important is not the number of hours spent helping others, but rather the purpose of your involvement. It is key to understand that the fire service is looking to hire people with a certain character and makeup.

The desire to be active in the community should come from within. It should be done for the right reasons, not because it will look good on an application. Frequently an applicant will inquire as to what would look best to a fire department, an involvement at the homeless shelter or becoming a little league coach. Unfortunately, these candidates are completely missing the point. You should not be doing these activities because they look good on a resume; rather, they are the right things to do.

Take a moment to reflect on the man or woman in your neighborhood who is involved in a positive way. He or she is always striving to make the world a better place and is a positive influence in society. Imagine if you put a firefighter badge and uniform on this person. It is not difficult to imagine all of the good he or she could do. If you can recognize what a good person he or she is, so can everyone else. This is the quality of the candidate we are looking for in the fire service.

Take an introspective look at yourself. When you tell people you are going to be a firefighter, do they say, "Really? I can't picture you as a firefighter." If they say, "Yes, I can see from the way you conduct yourself that it is a natural transition for you. In fact, I would like to introduce you to my friend who is a firefighter and perhaps he could help you out." If the latter is the reception you receive from your friends and acquaintances, you are probably on the right track.

*I do it to put a smile
on my master's face.*

The general public has a high degree of respect for firefighters. There is an expectation of how firefighters conduct themselves both on and off duty. If the people who know you best can see you in the role, this is probably a good barometer. If, on the other hand, they are surprised at your decision, it may be a good time to look in the mirror for a reality check.

This is not to say that fire departments are looking for community activists to become firefighters; rather, it helps them identify candidates with the attributes they are seeking. If it works well in the community, it works well for the fire department.

The fire service is about being unselfish. It is a way of life. Firefighters are continually volunteering on their days off to help out with things that are for the benefit of others. Much of our off-duty time is spent visiting sick children in the hospital, putting a roof on the home of an elderly widow who is on a fixed income, or pouring concrete at one of the fire stations.

Whatever the request, there is always somebody on the job with the expertise who is willing to lead the project. It is critical to have worker bees to help get the job done. If volunteering your time is not a priority to you before you get hired, what would convince the fire chief that you would be willing to volunteer your time after he or she hires you? Again, it is a way of life for most firefighters.

Some fire departments will bring up community involvement as part of the structured interview. Most candidates will struggle with this question. It may not be that they have nothing to say; rather, they have volunteered for the right reasons and are not seeking recognition. These candidates can usually respond well if they simply compose their answer. A candidate, however, who does nothing for the community will certainly not score well in this aspect of the interview.

Presenting your volunteer activity in an interview is difficult to do without boasting. It is akin to making a donation at church and then telling everyone how much money you gave. The fact that you boasted about it cancels out the good deed. It is more important to do the good deed for the right reasons than to try to get "credit" for doing so.

An opportune time to mention that you are involved in community activities would be during the initial question, "Tell us about yourself." Another place

would be to bring it up in the closing statement. You could present it by stating that you are aware that being a firefighter involves much more than working "eight to eight and then out the gate." There is an expectation that every firefighter will become involved in fire department or community activities on his or her days off.

I refer to this as leading the board. If they want to inquire as to what you meant, they will ask more about it. If they do not choose to pursue it, you can assume they know what you are talking about. Whichever the outcome, it is important that you are prepared to discuss your community involvement.

Community involvement comes from within. If you are struggling to "find" something to get involved in, then you are not getting the message. Get involved with groups, activities, or functions that you believe in. They are a statement about who you are. Your time is the most valuable gift. Use it wisely on things you believe in, not on what other people think you should.

Notes:

Notes:

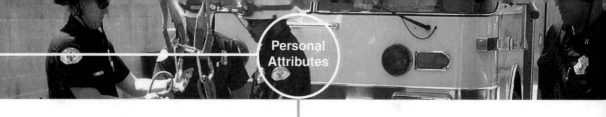

Would you hire you?

Tattoos

Tattoos have been around for hundreds of years and will be around for hundreds more. In today's society, tattoos are becoming more prevalent and in some segments of our culture, more accepted.

The fire department is a paramilitary organization. There are standards for dress code and for conduct both on and off duty. As a general rule, firefighters are a pretty conservative group, especially the older firefighters.

A candidate should understand that having a tattoo may affect his or her chances of getting hired. If the artwork is small and tasteful and not offensive to any racial group or society as a whole, it probably will not have any effect.

The more visible the tattoo, the greater the risk of affecting a person's chances of getting hired. The mindset of the fire chief will also affect the individual's chances. The more conservative the upper management, the less likely the department will hire a candidate with a tattoo. A department cannot legally disqualify a candidate for a tattoo, but it could certainly affect a candidate's chances.

I recently spoke with a candidate whose left arm was sleeved with artwork from the top of his shoulder to four inches past his elbow. He understood that a firefighter's job is to serve the citizens of the community in a professional manner. He was worried about how he would be perceived in the community.

He believed that his artwork was not inappropriate or lewd. He was not ashamed of it and planned to get more in the future. He was confident that the department would never find out about his tattoo until his first day on the job. He reasoned that they wouldn't see his tattoo during the physical ability test if he wore a long-sleeved shirt. "What do you think the department's reaction will be when they see my tattoo for the first time? Will they find some reason to fire me? If so, wouldn't that be some sort of discrimination?"

There is an underlying belief that a department will not see a tattoo before a candidate is hired; by then it is too late to terminate his or her employment. The rationale may be that during every contact with the department (interview, background investigation, polygraph and psychological exam), he or she will wear a suit or a long-sleeved shirt.

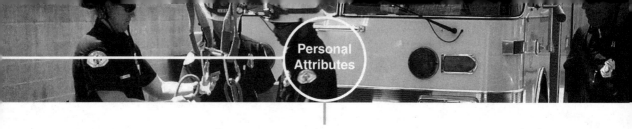

What would your co-workers say about you?

In reality, there is no "hiding" a tattoo. In addition to the physical ability exam, the department will conduct a thorough medical examination.

If a department decided to terminate a candidate, it would be very unlikely to be due to a tattoo. Hiring decisions are not made at the fire station level and the firefighters in the station do not know what is revealed or hidden during the testing process. If the department believed that a candidate were less than truthful, the department could make life difficult for him or her.

Having a tattoo will by no means prevent you from becoming a firefighter. Candidates see respected firefighters with large tattoos and believe they are acceptable. It is important to note that these firefighters probably got their large tattoos AFTER being hired. In other words, there is nothing the department can do about them.

If you are considering getting a tattoo, here is a barometer to use when contemplating your decision. Picture yourself reaching out to take my 67-year old mother's blood pressure. How is she going to feel when she sees your tattoos? Is she going to feel uncomfortable with you? If the answer is yes, your tattoo could affect your chances of getting hired.

Why is this important? It is the people in our community who encourage our city representatives to vote on a strong pay and benefit package for the firefighters. We spend a considerable amount of energy and effort to maintain a positive image in the community.

Some of our most competent and dedicated firefighters have ink on their bodies. It is even more popular since 9/11. I would trust my life to them. The key difference is that they already have a badge. If you are considering getting a tattoo and are wondering if it will affect you, the answer is a definite maybe. A wise candidate would postpone getting one in favor of waiting until he or she has completed probation. It is a risk versus benefit scenario. Why take a chance on affecting your future? If you still feel strongly about getting one, just be aware of the potential consequences of your decision.

Notes:

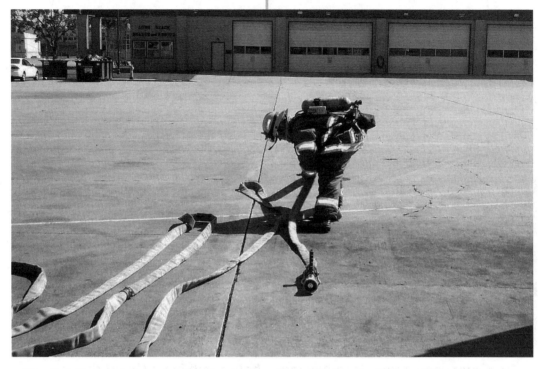

Firefighting can be physically demanding, particularly on older firefighters.

Age

Everyone has an opinion of age when it comes to hiring new firefighters. Some people feel that a younger candidate has a better chance of getting hired because, after all, the fire departments are looking to hire a candidate for the next 30 years.

If a fire department hires a 21 or 22-year-old, the department can train the recruit before he or she has a chance to develop "bad" habits. Furthermore, since the agency wants to get the most money for its training dollars, hiring a firefighter at a young age ensures that it will get at least 30 years of service out of him or her.

Younger candidates generally have fewer personal and financial obligations and are more likely to have the free time to pursue relevant education and

Older candidates generally mesh better with established crews.

training prior to being hired. This is highly prized by many departments, as they do not have to pay for it.

Younger firefighters are generally in better physical condition. They will do well in high impact areas of the community where the job is very physically demanding. In addition, they will usually work out in the station, which can be contagious to the other firefighters. Ultimately they may be the cause of the entire shift working out together.

Younger firefighters are often very concerned about eating properly and are more educated about nutrition. Quite commonly, older firefighters pay little attention to healthy eating in the fire station. A younger firefighter may educate the crew about eating turkey burgers instead of ground beef, or on the importance of taking vitamins.

Additionally, hiring younger firefighters minimizes the chances of hiring an employee with a pre-existing injury. It is true that a pre-employment medical exam will identify many of these injuries; however, with the implementation of the Americans With Disabilities Act, agencies are not failing nearly as many candidates as in years past. Since many candidates have successfully litigated and won a job, medical disqualifications have become less frequent.

The converse to these potential benefits is the fact that a younger candidate has spent the majority of his or her life at home with minimal responsibilities. Predictably, this will not be well received in a fire station. This is especially true since it is expected that the rookie is the one who makes sure all of the little things are done around the station. These are the same things that mom did at home for him or her.

Another factor when dealing with "younger" candidates is the fact that they are going to be living and working with mature (relatively speaking) adults. It can be difficult for a younger person to fit in with a group of older adults, especially firefighters.

Fitting in is difficult to begin with, especially when you consider that a respected member of the crew may have been moved to another station to make room for the new firefighter. The displaced crewmember probably contributed to the chemistry and cohesiveness of the crew and now an "outsider" has been assigned.

Maturity is an important quality for a young firefighter. Since he or she has usually led a sheltered life while in college or living at mom and dad's, it is likely that the rookie simply does not have extensive life experience. Imagine what you were like five years ago. How about 10 years ago? How much have your values and work ethic changed? I guarantee you are a different person. You have matured by virtue of your life experiences.

An older applicant, on the other hand, will usually fit in much better than a younger one. He or she has spent years in the work force learning what it takes to get along and has learned acceptable social behavior through "life experience."

Many departments prefer "older" candidates to younger ones. Since these departments are looking to hire firefighters with life experience, older candidates fit the bill. An older candidate will do whatever it takes to earn (and keep) the job. A candidate with more work experience may have a greater appreciation of his or her new job on the fire department.

Many older candidates have worked in a variety of difficult jobs. These range from roofing, carpentry, plastering or working behind a desk in corporate America. All of these jobs may include long hours, inadequate pay, little or no medical benefits, minimal flexibility, poor job security and, oftentimes, minimal job satisfaction.

A career in the fire service offers good pay and benefits, job security and retirement as well as job satisfaction. Hiring a more mature firefighter gives you a rookie who feels like he or she got a new lease on his or her employment life.

Older firefighters usually bring a lot to the job. If they have spent their lives working in the trades, they bring knowledge of plumbing, electrical and carpentry, as well as the skills of using various hand and power tools.

Most importantly, older firefighters generally fit in with the crew more easily than younger firefighters. Their life experience gives them a strong platform on which to base their career.

A candidate who is considering leaving an established job has a lot to lose. Add a mortgage payment, a spouse and a couple of children to the equation and this candidate has a lot on the line. The candidate is taking a pay cut, losing benefits and most importantly, losing job security. It is not likely that an employer will give an employee back his or her job after leaving it.

People who have a lot at stake make terrific employees. It doesn't matter how hard things get, he or she is going to have the drive to succeed. There is just too much to lose.

As you can see, there are benefits to hiring both younger and older candidates in the fire service. My personal belief is that most fire departments prefer to hire rookie firefighters who are in their late twenties to early thirties. Being married and owning a home strengthens their profile. Having a couple of children completes the equation.

This is not to say that candidates in their early 20's or early 40's will not be considered; they will simply have to demonstrate that they are the exception to the rule. It's up to the candidates to demonstrate that their personality traits, maturity and experience make them the best choice for the job. A fire department will consider much more than age when making a hiring decision.

Military Experience

Candidates who have served our country in the Armed Forces have a huge advantage over those who have not. It is generally believed that while military veterans may not have as many certificates and fire science units as other candidates (they were busy serving our country), they offer so much more.

There is no substitute for life experience. The personal growth a young man or woman experiences in the military is second to none. This growth is of course magnified depending on the assignments held. Many of those who join the military at a young age grow up very rapidly when put into dangerous situations.

Being assigned to the front line is not required to get "credit" for serving in the military. Fire departments realize that there are many support roles that require dedication and commitment. While there is only one person on the nozzle that puts out the fire, there are numerous other assignments that need to take place on the fire ground. It is important that a firefighter be willing to work in a support role for the good of the team.

The fire service is a para-military organization. Many of the common terms in the fire service, such as Captain and Lieutenant, were taken directly from the military. Words like code, honor, commitment, and integrity are clearly understood by those in the military. These qualities are also extremely important in the fire service, because firefighters are held to a higher standard than the average person in the community.

Men and women with military backgrounds are usually very mature, regardless of their age. They understand the need to get along with others, especially with people who come from backgrounds different from their own. Military people demonstrate respect for authority and understand the chain of command. The fire service operates on the same hierarchy principle as the military.

Physical fitness is emphasized in the military. As a result, military men and women are usually in very good shape. This is extremely important to the fire service, because the number one reason entry-level candidates fail out of the academy is due to poor physical fitness. In addition, a physically fit firefighter will miss less time due to injury than a firefighter who is not fit. Military personnel have been taught the importance of a life-long physical fitness program and the importance of proper diet. These good habits will be shared with the firefighters in the station.

Military men and women are used to working in a structured environment. They understand commitment and the need to work until the job is completed. They are used to working for long periods of time in less than ideal conditions. They understand the importance of doing something right the first time. Similar to the fire department, people's lives are impacted if things are not kept in a constant state of operational readiness. Firefighters must check out their equipment each and every day. They must know the intricacies of each tool kept on the engine or truck. Training and continuing education are essential to the fire service. It is imperative that firefighters are able to work unsupervised; completion of a job or task is a reflection of them.

Getting along in the fire station is critically important to being successful in the fire service. Courtesy to fellow firefighters is critical. Cleaning up after oneself is expected. This is one of the first things learned in Basic Training in the military.

One of the strengths found in military men and women, however, is also commonly a cause of strife during their probationary year. People who have earned rank in the military are used to giving orders. As a rookie firefighter you are expected to take orders, not give them. Humility is an extremely important quality to possess as a rookie firefighter.

Rookie firefighters who have spent time in the military are often older than the average candidate. It is not uncommon for an older probationary firefighter to be working under the tutelage of a much younger senior firefighter, engineer, or even lieutenant or captain. If the rookie firefighter does not have the proper mind set, he or she will be in for a difficult probationary year.

If you are still in the military and are interested in a career in the fire service, it is important that you start making provisions NOW. If possible, put yourself in a position to get fire service-related training such as Medic or Corpsman. Hazardous Materials and firefighter training will also be beneficial. Lastly, work on general education courses so you can earn your Associates degree. Start taking online classes NOW.

Do not be intimidated by all of the candidates who have every certification under the sun. They were able to obtain these as full-time students while you were busy fulfilling your commitment to the American people.

A candidate who is an EMT, possesses related experience as a reserve or volunteer firefighter, and is active taking fire science courses is usually at the top of his or her game. Get your qualifications, learn how to take a fire department interview, and earn your badge.

Notes:

Notes:

What to Expect

The following is an excerpt from my book, "Smoke Your Firefighter Interview." Although it may be a review for those who have already read the book, I felt it important enough to include here.

The First Day

The following article is being reprinted with the permission of the author, Captain David Shetland of the Long Beach Fire Department.

The fire service is going through unprecedented turnover due to the Baby Boom generation retiring. Hiring is at an all-time high.

The downside is, we are losing an entire generation of experienced firefighters. Most had military experience; as a result, their work ethic, motivation, and esprit d'corps served as an excellent example for rookies.

With these personnel gone, firefighters with only three or four years experience may be thrust into the unfamiliar position of being the most experienced person on the job – the "Bull Firefighter." As such, he or she must serve as a role model to new hires.

This information is based on years of instruction that was given to me several years ago by my senior firefighter. I have shared it dozens of times with other firefighters, and it has served them well. Now I will pass it on to you.

I'll always remember what the drill instructor told me on the last day of academy: "You only have one chance to make a first impression." Since then, I've done a lot of thinking about ways to make a positive first impression – hopefully, one that will last your entire career.

After receiving your initial assignment, you should probably visit the fire station before your first shift. This offers several advantages over showing up the morning of your first day of work. Among them:

- Locating the fire station ahead of time ensures you won't get there late on your first day because you got lost.
- It's advantageous to become familiar with the firehouse in advance. Where do you park? Is there a gate and, if so, will you need a key? Will

you be issued a station key? Where should you put your turnouts, and what time should you arrive?

- You can talk with the other rookie(s) to learn the morning routine and find out what duties need to be completed before line-up. For example, what time is the flag raised? When is the paper brought in and coffee started? Find out any other special information that may pertain to that particular station, and ask for a tour.

- Don't forget the importance of becoming familiar with the apparatus on which you will be riding. You are going to be a vital part of that apparatus and its crew. Know it; live it; learn it.

When you show up on this reconnaissance mission, be sure to bring ice cream or another treat with you. This will go a long way with the crew.

When your actual first day of work arrives, it is a good idea to show up in uniform (and carrying donuts). Be nice and early.

The absolute first priority is to find out which apparatus you will be riding on and where. Once you have learned this information, relieve the person in that seat and check over the apparatus thoroughly.

My suggestion is, once you have put your turnouts on the rig, immediately check your breathing apparatus and all its functions. If someone should ask you the pressure in the bottle, you should know it!

Now go through the rest of the rig with the same resolve. Your life, as well as the other members of your crew and the citizens you are sworn to protect, depends on this equipment.

Here is something else to consider: If your fire station has multiple rigs, you should know them all. At a moment's notice, you just might be temporarily assigned to another piece of apparatus.

Once you have completed these tasks, make sure to get right to your daily duties. Finish them expeditiously.

When the captain calls line-up, be the first one in the kitchen. While you're waiting for the other members to arrive, straighten up the kitchen and start serving coffee. Be the last one seated, and provide service throughout the line-up.

It just doesn't look good for you to stand around while a senior member of the crew serves coffee. We've all had our "day in the sun," and now it's your turn.

Once line-up is done, start working on the weekly duty. Don't wait for someone to tell you what to do.

If at any time a run comes in, make sure you are the first one on the rig. Write down the address and the nature of the call.

It's not uncommon for other crewmembers to say, "Where are we going, and what is it for?" This is especially true if the department doesn't have MDT's. The point is, if they ask and you have the information, it's Brownie Points for you.

When you have completed your daily and weekly duties, what is next? This is a good time to ask other crewmembers if they need assistance with anything. If they don't, it's time to go through the rigs again and again.

What kind of impression will you make if a piece of equipment is requested and you don't know where it is? You know the answer to that! I can't stress enough the importance of not only knowing where everything is, but what it is and how to use it.

As a rookie, you're under the microscope. In other words, you're going to be scrutinized throughout your probation.

The following are some general guidelines for not only making a good first impression, but also creating a reputation you can be proud of your entire career:

- Introduce yourself to everyone. Don't wait to be asked who you are.
- Don't be afraid of mistakes! If you're not making mistakes, you're not learning.
- Never give up your tools. If someone asks to borrow them, inform the person that as soon as you have completed your initial task you will be happy to assist him or her.
- Always answer the phone.
- Show initiative – do things without being told.
- Do one extra task per day for your station or apparatus.
- Be seen but not heard.

What
to
Expect

- If you're not sure if you should be in full PPE, err on the side of caution and fully suit up.

- Be the last one to bed and the first to rise.

- Prepare completely for drills. Remember, the rest of the crew has heard the drill numerous times. Do everything possible to give them some new information they may not have heard before.

- Always give 110%. You want others to tell you to slow down, not speed up.

- Remember, if they want your opinion, they will give it to you!

A career in the fire service is a privilege – so no complaining about being interrupted during dinner or after you go to sleep. When you are given what you feel is a tough or crummy assignment, remember that you don't "have" to do it, you "get" to do it. Never forget that!

Being a rookie is not an easy task. The fire service is filled with old traditions and quirky nuances, but if you start with these simple guidelines, you're sure to create outstanding habits and make a terrific first impression. Believe it or not, you'll probably look back on your probation as the greatest time of your life!

Raising the flag with pride and honor.

Rookie Life

The probationary period is usually the first year of employment. Although one year is customary, the time may be as long as 18 months or as short as six months. The time spent in the academy may or may not count toward the probationary period. At the end of the probationary period the department has the option to pass or terminate the recruit firefighter.

By the onset of the probationary period, the department has invested a lot of time, energy and money in the employee in the form of a medical exam, polygraph test, psychological exam and thorough background check. They have also provided the employee with training, either on-the-job or in the form of a formal fire academy. During the course of the probationary period, each recruit is expected to complete a series of written and practical exams. The

Everyone goes home safe.
— Fire Chief Charlie Hines

written exams are based on reading volumes of the department's policies and procedures, as well as the operational manuals. The practical exams are based on the fundamentals he or she learned in the academy, combined with real life experience gained while working as a recruit firefighter.

While the department has invested a lot of time and money in each recruit, this in no way means there is a guarantee of success. At the end of the probationary period, the department will determine if the recruit is worthy of being promoted to full-time permanent status. Although the vast majority of recruits do make the final step, it is not unheard of to be terminated on the last day of probation.

The best analogy for the probationary period is that of the department loaning you a temporary badge. At the end of the specified time frame your captain (with input from the crew) will decide if you get to keep it. If you have proven yourself "worthy" and have gotten along with your crew, the decision is easy. If not, the decision may not go in your favor.

The following was written by an anonymous rookie firefighter who recently completed a fire academy at a large, extremely traditional fire department in Southern California.

My typical day as a rookie firefighter starts off at 4:30 a.m., waking before the sun comes up. I rehearse my drill for the day prior to leaving for work.

5:10 a.m. I arrive at the station and open the gate.

5:15 a.m. I enter the station and put up the first pot of coffee. I proceed to the bathroom and change into my fire department uniform. I return to the kitchen and make the second pot of coffee. I continue to the apparatus floor to get my turnouts in order on the engine or truck. I progress to the captain's office where I check the journal to see yesterday's activities as well as check the roster to see who I will be working with for the day. Lastly, I check the "new material" for any pertinent information pertaining to the department or today's activities.

5:35 a.m. I put up the American flag and gather the newspaper and return to the kitchen and spread it out, section by section, on the table. I then empty the dishwasher.

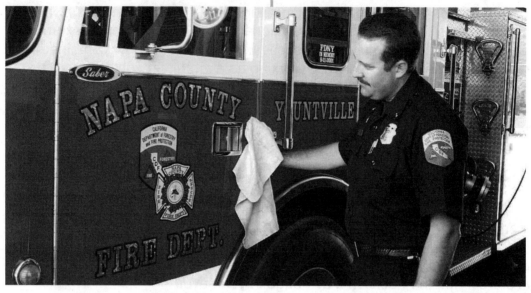

A rookie firefighter's job is never complete.

6:15 a.m. My crewmembers begin to arrive as the off-going crew begins to wake up. I make it a point to say "good morning" to each and every member. If I haven't met someone, I make it a point to introduce myself and not wait to be asked who I am.

6:25 a.m. I find the other rookie so we can practice throwing every single ladder as well as practice donning our self-contained breathing apparatus (SCBA) for time. Periodically, between ladders and SCBA practice, I will return to the kitchen to make more coffee.

7:15 a.m. I practice my daily drill with one of the senior firefighters. He or she will help me make sure my drill is prepared and ready for the rest of the crew.

7:45 a.m. I proceed to the kitchen to prepare for the shift's official line-up and make more coffee. I also clean up the mess made by the senior firefighters while making breakfast.

8:00 a.m. I line up in the kitchen with all of the members of my shift. We go over the itinerary for the day and discuss any new material and departmental happenings.

*Walking on roofs allows firefighters to
become proficient with vertical ventilation.*

8:30 a.m. I begin the housework details. I always make it a point to be the first one cleaning the bathrooms with my scrubber and bleach/Comet mixture. I have learned that instead of flushing the toilet once clean, leave the soapy water in the bowl. This shows your crewmembers that the toilet has been cleaned.

9:30 a.m. My crewmembers begin their physical fitness routine. The other rookie and I are busy throwing ladders, doing our daily/weekly maintenance checks and practicing our daily drill.

10:30 a.m. We are en route to the store to shop for lunch and dinner. While at the market I will throw ladders, give on-the-spot drills on equipment, walking on roofs or doing something practical.

11:30 a.m. I help the cook and set the table for lunch.

12:00 p.m. Lunchtime! I am always the last to gather my plate unless otherwise ordered. I usually take the smallest portion to make sure

there is enough for everyone. Even though I am usually the last to sit down, I am always the first one to get up and get into the dishes. I eat so quickly that most of the time I don't even taste the food. I jump into the dishes until the cook calls for a "game" to decide who will officially be stuck in the dishes. This usually entails some type of dice or card game. I intentionally lose because it would not be correct to have the rookie at the table while the captain is in the suds.

1 p.m. I will help the engineer or other senior firefighter with projects that need to be completed around the station or apparatus.

2 p.m. I will give my drill in front of the 12 members of my crew. I have presented it at least three times before, but now the pressure is on. As you can imagine, each one of the firefighters has a tremendous amount of knowledge about the subject that I could never have learned in a book. It can be a bloodbath if I am not prepared. I find that if I take the time to do my research, I usually can come out of it alive. If not, it can be very difficult.

3 p.m. I pull out the tool that I have been assigned for my drill on the following shift and begin reacquainting myself with it. I research the tool in the technical journals and begin to gather my notes. When I get home, I will research on the internet for more information.

4:30 p.m. I clean the kitchen from the afternoon's snacking. I help the cook prepare for the dinner meal.

5:30 p.m. I take down the flag and double check that the gate for the parking lot is locked to maintain security for the firefighters' private vehicles.

6 p.m. Same routine as lunch. I am the last to sit down and the first to be in the suds.

7 p.m. I help the engineers wash and chamois down the apparatus.

8 p.m. I will pull out another tool and begin to learn it. I will pull a ladder off the engine or truck and throw it, read the policies and procedures, or prepare for my drill next shift.

10 p.m. I do a final cleanup around the station, picking up any residual trash, doing the dishes again and doing a final inventory of the engine or truck.

1:30 a.m. I finally go to sleep when the last member of my crew has gone to bed.

5:30 a.m. I wake up before the rest of my crew, put on my uniform and make coffee. I open the gate, get the newspaper and make sure the kitchen is clean.

8:00 a.m. I change out of my uniform and leave the station after the last member of my crew leaves.

This is just a rough baseline of what to expect as a rookie firefighter. It is important to note that this does not include running emergency responses and all of the on-the-spot questions that barrage you during the course of the day.

Notes:

Notes:

Fire Academy

The following was written by an anonymous rookie firefighter shortly after being hired by a major Southern California fire department.

I recently graduated from a tower this past spring/summer where 50 started but only 30 graduated. This is almost a 50% failure rate. I can only share my experiences of what I saw. If you talk to other people, they may have keyed into different things.

Poor Attitude:

1. Igmr's (I got mine) – if you have this mind set the instructors will quickly identify you as someone who is not a team player.

2. Be a listener, not a teacher. If you know something, share it with your classmates during lunchtime. Don't suggest something to an instructor about a trick you learned as a fire explorer or as a firefighter from

another fire department. Remember, you are trying to pass the tests (manipulative and academic) the "tower" way, not the "field" way.

3. Keep your ears open and your mouth shut. Only chitchat with your buddies at lunchtime. Don't join into conversations that shouldn't be going on in the first place.

4. Don't talk badly about your instructors or your fellow cadets.

5. Don't make excuses. If you screw up, don't apologize; just move on. Most importantly don't make the same mistake twice.

6. Don't go out with your buddies on weekends to "take a break," because that's how people get into trouble. DUI's, fights and public intoxication are a sure way to get dismissed from the academy.

7. Do not brown nose your instructor. They are not your friends, nor will they ever want to be. Show respect and you will do fine.

8. Remember you are there for a badge, not to gain friends. Keep the non-essential talk for after you leave the drill tower grounds.

9. Support your fellow cadets as much as you would want to be supported. You will not make it through without their help and vice versa.

Physical Conditioning:

The first three weeks were the most difficult. It appeared they wanted to weed out the weaker candidates. We had 13 people quit in the first week and a half, many of these in the first two days.

The physical ability test is not even close to the exertion you will go through in the tower. If you barely pass the ability test, you are in trouble. Each day you will go home sore, bruised and strained. Due to the fast pace, your body does not have a chance to recover from one day to the next. The better your physical condition, the greater the chance your body can adapt to the rigorous training. It is imperative to be in the best shape possible. If you aren't, you are going to get hurt.

Mental Conditioning:

After the first four weeks of our 14-week academy, it started sinking in that we were going to be here for a while. It's mentally draining. You have to stay focused or you will never make it.

It is extremely stressful to prepare for a manipulative exam knowing that if you don't perform you will lose your job. Everyone in the academy had to perform an evolution a second time knowing that this was his or her last and final opportunity. I guarantee it will happen to anyone who enters an academy.

Being able to perform under pressure is critical. Remember, you are your own worst enemy.

Academic:

You will be exposed to information about a myriad of different topics while in the academy. You are expected to know every piece of information that has been presented. You will be tested on it weekly, sometimes daily.

People failed out of my academy for a variety of reasons. Probably the main reason was poor physical conditioning. Even those who survived the first 10 days had physical conditioning issues. It was apparent who was struggling. When you are tired and run down, you don't think clearly. This leads to mistakes, which in turn lead to bringing attention to yourself. Ultimately, you find yourself fighting for your job.

There are many things you can do to enhance your opportunity for success in the academy. First and foremost, maintain top physical conditioning. The better shape you are in, the better your chances of avoiding injury and making unnecessary mistakes.

Secondly, put yourself through a fire academy at the local community college. The more familiar you are with ladders, hose and SCBA's, the better your chances of being successful in the academy.

The academy is extremely fast-paced. Those who did not have previous experience to draw from definitely had a more difficult time. Fortunately I had been through a basic fire academy. I have to admit that the academy at the community college, although at the time seemed hard, was like a day at Disneyland compared to the fire department's academy.

Learn how to study before you enter the academy. Find a place where you can sit down and get away from the world and immerse yourself in the books. Set it up beforehand; don't wait until you start the academy to figure out where you are going to study.

Form study groups early. Take a look around and try to identify who appears to be focused on making it through. There is no doubt that there is a benefit to having someone to bounce questions off. He or she may interpret the reading material differently than you and key into something you may have

misinterpreted. In addition, he or she will pick you up when you are struggling and vice versa.

Take fire science courses prior to entering the academy. The more background and exposure you have to the fire service, the better you will fare. Remember, each night you will be assigned a ton of reading. You are physically exhausted after being on the grinder all day long. It is difficult to maintain concentration to sit and study for a written exam the next day. The more information you have before entering the academy, the easier the material is to digest in a shorter time frame.

Completing the academy is one of the most challenging things you will ever go through. The more you can stack the deck in your favor, the better the chances of making it through. Don't take it lightly. The work is just beginning.

Notes:

*You can accomplish anything
you put your mind to.*

Paramedic School

Tom Rollins, a graduate of the Daniel Freeman UCLA Paramedic program, wrote the following article. Rollins has agreed to share his experiences with future firefighters, so that they may make an informed decision when deciding to go to paramedic school.

The decision to go to Paramedic School is one that should not be taken lightly. It will be one of the most challenging periods of your life, and to jump into it without serious thought, preparation and planning could spell disaster to your goal of becoming a firefighter. With that said, it can also be one of the most rewarding periods of your career in the fire service.

I thought I was ready for paramedic school when I applied a few years back. I was working as a reserve firefighter in a very busy part of Los Angeles County and ran multiple 911 calls every shift. I had already attained an A.S. degree in Fire Science at a local college, put myself through a fire academy and was working one 24-hour shift a week as the fourth person on a very busy engine company.

The final straw was when I spent two nights in line to get a job application for a local fire department. The first night of the line-up, a battalion chief walked the entire line with a handful of applications, handing them out to anyone who could show him a paramedic card. The rest of us stood in line another 30 hours. By the time I got to the front of the line, they had run out of applications. Instead of taking tests with 2000 of my closest friends, I told myself that this was going to be the last time I slept on a sidewalk just to see the job go to a paramedic. I was convinced that the next natural progression to becoming a firefighter was attending paramedic school.

I was overly confident in my abilities as a student because I had sailed through the courses toward my college degree with very little effort. In addition, EMT classes were a breeze for me. I had completed some upper division college classes at the state university level in Biology and Pre-Dentistry (I had aspirations of becoming a dentist before I realized my true calling was in the fire service). I was actively instructing first aid and CPR classes for about eight years prior to applying. Little did I know what lay in store for me. I came

to find out that my story was quite typical of my future classmates as well.

This brings up a good point. Are you going to paramedic school just to be able to check a box on your job application? Or are you going because you have a real desire to learn more about pre-hospital care? I saw many "box checkers" fail out of the program because the effort it took far exceeded the desire to have a "P" nailed onto the end of their EMT card. If you are hired as a firefighter/paramedic, you will be expected to work as a medic probably for quite some time. If you don't like being a medic to begin with, it's bound to show. You are going to be a very unhappy person who is being scrutinized on every call. It's hard enough to be a rookie firefighter without the pressure of being a paramedic at the same time. So give some serious thought about jumping into paramedic school if you're not ready or not really willing.

To get into a paramedic school you have to meet some basic requirements. Since they vary greatly from school to school and state to state, I won't go into them here. After these requirements are met, a typical program will require you to take a basic EMT-1 level test. This is the first weeding out process you will encounter. My school had a minimum acceptance level of 85% to go on in the process. After that you are invited to take basic math, reading, writing and comprehension tests. The third step is an oral interview similar to a fire department oral board, where they ask you a few situational questions and your reasons for wanting to become a paramedic.

If you are accepted, you go in for an orientation and receive your books. I suggest bringing a large backpack and parking as close to the front door as possible. You will most likely be assigned some study material before the first day of class. The first morning you walk in, expect a quiz. The instructors are testing your ability to follow directions. If you don't score well on your first quiz after having weeks to prepare, the instructors will have a nice one-sided conversation with you in which you do most of the listening. "How are you going to keep up throughout the program with only hours of study time instead of weeks and score above 80% on every quiz?"

On the first day of instruction a doctor spoke to our class and told us that he demanded excellence in us. We were going to learn at a pace that was similar to a first year medical student and would be expected to perform at that level as well. If anyone didn't think that they could hack it, he invited them to quietly leave at the end of his address, no questions asked and receive a full refund of their tuition. He said that for the rest of our lives (yes that's right, the rest of our lives) we would remember paramedic school and what we had to endure to graduate. And I can assure you, truer words were never spoken. You will always remember your time spent in medic school.

It's not that the subject matter is all that hard; if given two years to prepare and study, I'm sure that most people could graduate. The problem is that many programs teach the course in six months. Talk about putting a ten-pound chicken in a five-pound bag! So as you can probably see, medic school is all about mastering a vast amount of information in a short amount of time.

A paramedic program is typically broken up into three phases: didactic (classroom), clinical and field internship phase. I was tested every day with written quizzes or skill stations. The minimum passing level in my program was 80%. No grading curves, no excuses, no missing classes and no sleep. OK, I'm joking. I was able to sleep one to three hours on most nights during the classroom phase.

The following suggestions will help you prepare for and get the most out of each phase of school. Through planning and preparation you can increase your chances of graduating and getting your paramedic license.

Before even applying to paramedic school, I suggest you take a semester course of Anatomy & Physiology (lecture and lab) at a local community college. In fact, many programs are starting to include this as a prerequisite. This is a good foundation class and you should work hard in it. Keep in mind that most paramedic schools demand at least 80% to pass; you should be in the upper 90% in every pre-paramedic course you take.

Next is a basic EKG (Electrocardiogram) course. You don't need to master 12 lead EKG's yet, but it wouldn't hurt. Basic EKG courses are taught in three days or less. Know every cardiac rhythm taught to you and know it well. Be able to read a rhythm strip at a glance, not with five minutes of debate with calipers in one hand and flash cards in the other. Your field internship instructors will expect you to know this cold.

A course in medical terminology, ACLS (Advanced Cardiac Life Support), PALS (Pediatric Advanced Life Support) and any other course that you can think of that will expose you to pre-hospital medical training are all feathers

Volunteering to be a victim is a great way to learn about paramedic shool

in your cap to help you get into paramedic school and succeed. If your basic math skills aren't what they should be, make sure you do whatever you must to get them there. Pharmacology is all about fractions, decimals and conversions factors. Study up on your metric system, paying particular attention to volume and mass measurements like milligrams and cubic centimeters.

Many programs are now offering a paramedic prep course to help students be successful. All these courses count when it comes to admission time. Many people apply to paramedic school and the competition to get in is growing every day. If you can show that you are better prepared than the next person, chances are you will get the slot and not end up on the ever-growing waiting list. The school wants you to succeed.

Another good way to prepare for a paramedic program that is often overlooked is to become the best EMT-1 you can be. Let the paramedics that you work with know you want to prepare yourself for paramedic school. I bet they will let you do some patient assessments and run through some patient simulations. I spoke to a paramedic program instructor who said patient assessment skills are severely lacking in his new students. Your field internship will go much more smoothly if you have actually done a few patient assessments and not merely acted like an IV pole on all of your 911 calls. The key is to get in there and get the experience.

Some fire departments insist that their rookies do ALL the primary assessments prior to the paramedics taking over with the advanced stuff, so you might as well get your hands dirty. When the medics are doing something you don't understand, ask them after the call why they chose that certain treatment. Get to such a level of competence that you can predict what medication is going to be administered and why. You will be doing the same thing in the near future, so pay attention.

A huge leg up in preparing for paramedic school is knowing your drugs. There can be over 100 pre-hospital drugs to learn and most of it is just rote memorization. You don't have to know what the drug does to memorize its dosage, indications and contraindications. By knowing this prior to the first day of class, you will buy yourself some much needed time to study other subjects (or sleep) while everyone else is struggling with pharmacology. Visit any paramedic school and they will gladly sell you the most current pharmacology handbook.

You can also use this visit as important face time. Speak with an instructor or sit in on a lecture if you can. Talk with some of the zombie-like students and ask them how you could be better prepared for your class. When it comes time to take the oral interview, you may see a familiar face on the other side of the table.

Another way to get experience is to volunteer for some of the simulation stations at the school to which you are applying. Often the school is looking for mock patients for the current class. This will give you an opportunity to see how the students are tested and what constitutes a pass or fail in a skills station. You would be surprised how often the program needs volunteers. And again, this is more face time for you to talk with the instructors and get information that other applicants will not have. The competition for that spot in the next class is high; all the above things will hopefully tilt the odds in your favor.

When you do get that acceptance letter you need to prepare both mentally

and financially. In my paramedic class we lost right around 40% of the students by graduation. A lot of these students were trying to work a job while in the program. If at all possible, *do not try to work while in paramedic school* if you are in a full-time, six to eight month program.

The didactic period will last from 9 to 16 weeks depending on the school you attend. This is usually a Monday through Friday 8 a.m. to 5 p.m. schedule. But that is just class time. Next you have to drive home and eat and then study. You must explain to your family and friends that you are unavailable to do anything for the next 9 to 16 weeks. No nights or weekends off; these will be spent studying. All free time will be spent studying material, working on your assigned project, or preparing for the next test. All of your loved ones will have to excuse you from any other responsibilities during this time. If you fill your plate with anything other than paramedic school, you will most likely fail. Class failure rates in the 30% to 40% range are not uncommon.

The next step in the program is called the clinical phase, during which you will spend about 180 hours in a busy hospital emergency room. A lot of this time will be spent starting IV's and generally angering your patients in the process. You will also get the chance to practice your patient assessment skills. Do as many assessments as you can. If you don't show any aptitude, the nurses will be more than happy to forget about you and move on to someone who is more interested in becoming a better paramedic intern.

Some people use the clinical phase to coast and relax before going out to the field internship. Don't become one of these people. When a paramedic comes in with a patient, listen to how he or she gives report. Ask the medics what drugs were administered in the field and why they gave them. Listen to incoming 911 calls if you are in a base station hospital and ask the nurses for pointers on how to talk on the radio. When a patient needs to be intubated, make sure you volunteer for it. Sometimes the doctors forget that this is in your scope of practice and will do it themselves, but they might let you do it if you speak up. I got to do a number of intubations this way. When a trauma or full arrest patient comes in, make sure you're not off doing something else less important. Get in there and practice your mega code skills and listen to how others run the code. Take notes as soon as possible after the code and write down everything you didn't understand or were confused about. Go over the code with the nurse and ask questions.

The field internship is the 3rd and probably most challenging portion of your schooling. You will be riding out with a busy paramedic unit for at least 20-25 shifts of 24 hours in length. Many paramedic programs have a difficult time placing students because the paramedic preceptors are taking you on as a favor to the school and/or program in general. Keep in mind that your preceptors get no extra pay for accepting you into their lives and they are doing *you a favor!* All they get out of the deal is more paperwork, more hassle and more headaches. In return they get to mentor new paramedics and this is why they really do it. If you are lucky enough to be able to do your internship at a municipal fire department where you will be working with highly experienced and knowledgeable paramedics, the whole crew is taking you under their collective wing. Treat this opportunity as you would a rookie firefighter position, because you may be one at this department if everything goes well.

I highly advise trying to find your own internship before you start paramedic school. This way you can get a good internship in a busy fire station and get the most out of your time. It's better to be graded on ten calls per shift than two. Imagine the upper hand you would have on an oral board if you were a paramedic intern at that department. Need I say more?

No paramedic internship goes perfectly. Your preceptors don't expect perfection on your first shift except in one area – pharmacology. You may have remembered that I mentioned this at the beginning of the chapter. In your first shift it is common for the preceptors to question you on drug dosages, indications and contraindications to get a feel for how much you have prepared yourself for the field internship. If you can't rattle off all your dosages like a 4th grader recites the alphabet, a whole new can of worms may be opened up and they will start questioning everything you learned. Not a good way to make that first impression. Know your drugs, know your drugs, know your drugs…there, I said it three times.

Keep a positive attitude at all times when you are in your field internship. When you mess up a call, and you will, learn from your mistakes and move on. Visualize your next call going perfectly. When your preceptors tell you to change or add something to your patient assessments, do so immediately. Never argue with your preceptors or disagree with them in the middle of a call. There will be plenty of time to discuss the run on the way back to the station. During your internship you should never tell your preceptors how you will deal

with patients in the future; just keep your comments to yourself and try to learn as much as possible in your 20 shifts.

The assessment form that your preceptors fill out every shift has a section in it that grades your ability to take instruction and criticism. You will be amazed that what seems like a stupid idea on shift number five makes perfect sense on shift 18. A sure way to fail is to argue, disagree, or not follow instructions. The valedictorian of my class failed his field internship because of his inability to take instruction.

Make sure the rig you ride on is the cleanest, most well-stocked rig in the city. When the next shift comes in and checks out the rig, everything should be fully stocked and in its place. The scope is clean and shining with a new roll of printer paper. The EKG patches are overflowing out of the pocket and the leads are wiped clean. The drug cabinet has no expired drugs lying around in the dark hidden corners. (A nasty little trick preceptors like to do is hide an expired drug in the bottom of the meds box.) The drug box is scrubbed to a shine and all the brass on the clasps is polished.

When you come back from a call, restock any item that was used on the patient. Never let the scope batteries run low, or an audible alarm will sound off telling everyone on the next call that you are slacking. Empty out a compartment and wipe it clean, throw out any trash and put everything back in its proper place. Make sure you don't leave any oxygen bottles empty. If you find any medical equipment that you don't understand and can't give a drill on with ten minutes notice, ask your preceptors.

When you have nothing to do, your nose is in the books. Every firefighter who sees that rig will know a paramedic trainee is on it because it will be shining when it comes down the street and every paramedic who works on the rig will not have to lift a finger when it comes to restocking it. If you work hard at everything you do in your 20 or so shifts, you are showing everyone how much you want to be there and do well. A good trainee acts a lot like a good firefighter rookie.

Station Drills

A firefighter is responsible for knowing a tremendous amount of information, ranging from knowledge of his or her "first in" response district, application of a traction splint, and patient assessments, to hose and ladder evolutions, just to name a few. During the probationary period it is up to the captain (and crew) to ensure that a rookie firefighter is proficient in all aspects of the job.

It is impossible to make sure a firefighter is competent in all aspects of the job, since many of the duties required do not present themselves on a regular basis. For example, a firefighter may only need to set up a portable monitor or throw up a 35-foot extension ladder once every couple of years. It is still imperative that he or she is proficient in the skill, as time is of the essence when it needs to be done.

Much of what is learned and demonstrated in the fire station is conveyed during station drills. These range from an impromptu question about the construction of the roof of the grocery store while you are shopping, to a formal presentation with computer-generated handouts.

The crew gathers around the kitchen or day room and a member reviews a piece of equipment, outlines a policy or gives a refresher course on a particular topic. On a seasoned crew it is usually the captain who acts as the teacher and coach; however, another member of the crew may deliver the drill. If there is a probationary firefighter on the crew, rest assured he or she will be assigned the task.

As a general rule, a rookie firefighter will be assigned a different drill to put on each shift. He or she may be expected to research a piece of equipment and provide all of the facts pertaining to it. All methods of how to use it must be covered, as well as alternative means. If the primary method fails, what is the back-up way to make the piece of equipment function the way you want it to? Lastly, what is the proper maintenance schedule for it?

Researching a piece of equipment is done through the manufacturer's specifications, the Internet and through the knowledge of your senior firefighters. It is difficult to shine when you realize that you are putting on a comprehensive drill on a subject or piece of equipment that the crew is already proficient on. In fact, there is a good chance that they are subject matter experts. It is critical

that you do your homework and know your information. You will not be able to bluff your way through.

Since there has been such a great number of new firefighters hired in the last several years, the amount of in-house training has escalated. As each subsequent group went through their probationary period, the expectation was raised to a new level. Because firefighters are a competitive group by nature (they never would have gotten the job if they weren't), each group of new firefighters tries to out do the previous one when it comes to the thoroughness of a station drill. Since there are only so many topics, they are repeated with each subsequent drill. It becomes a game to try to stump the rookie with a question during his or her drill.

Since the expectation of the station drills is so high, nothing less than excellence will be accepted. It is not uncommon to spend several hours preparing handouts and researching for a 15-minute presentation to your peers.

Being a good public speaker is a great asset. If a rookie has a fear of speaking in front of a group, he or she is at a disadvantage, as the new firefighters are usually assigned to the busy downtown stations. Many of these large stations have 10-15 firefighters assigned to the crew. As you can imagine, this can be very intimidating. Any type of teaching experience will greatly enhance a rookie's ability to conduct a professional, well-organized drill.

Below are a few examples of station drills, along with the computer-generated handouts completed by Firefighter Darren Strecker.

PORTABLE FIRE EXTINGUISHERS

INTRODUCTION

Portable fire extinguishers are available to the firefighter from responding apparatus or from public and commercial buildings equipped with various types of extinguishers. The trained firefighter learns to depend on the Fire Department's own equipment, but one should also become familiar with other commercial type fire extinguishers. One should know their operation and limitations on various classes of fires, both at the time of emergency use and for inspection purposes.

Extinguisher Rating System

Class A: *Ordinary combustibles*

**Ordinary
Combustibles**

- Rated from 1-A thru 40-A.
- The number is indicative of the relative fire extinguishing potential of that extinguisher (by size or contents).
- A "2-A" is capable of extinguishing a wood crib two cubic feet in size.

Class B: *Flammable liquids*

**Flammable
Liquids**

- Rated from 1-B thru 640-B.
- The number is indicative of the square-foot area of deep-layer flammable liquid fire (1/4" or more) that can be extinguished.
- A "20-B" is capable of extinguishing 20 square feet of flammable liquid.

Class C: *Energized electrical*

**Electrical
Equipment**

- No numeral is used since Class C fires are either Class A or Class B fires involving energized electrical wiring and equipment.
- Testing for Class C extinguishers focuses on the conductivity of the extinguishing agent as well as the hose and nozzle combination.

Class D: *Combustible metals*

**Combustible
Metals**

- No numeral is used since there are so many metals, which would each require different ratings.
- The relative effectiveness of those extinguishers for use on specific combustible metal fires is detailed on the extinguisher faceplate.

USING A PORTABLE FIRE EXTINGUISHER

Follow the instructions and the acronym **P.A.S.S.**

P – Pull the safety pin.

A – Aim at base of fire.

S – Squeeze the hand held trigger.

S – Sweep the target.

- Approach from windward side.
- Apply agent from a point where it reaches, but does not disturb the fuel.

COMMON TYPES OF PORTABLE EXTINGUISHERS

2 ½ Gallon Pressurized Water Extinguisher/ 2A

Use:

- Class A fires.
- Stainless steel construction.
- Shell tested to 600 psi.
- Delivers 30'-40' stream for 30-60 seconds.
- 1-2 oz. of Fine Water or Tergitol increases the penetrating characteristics of water.

Testing:

- Hydrotest every five years.

After use:

- Add water to mark on shell.
- Add 1-2 oz. of Fine Water or Tergitol.
- Pressurize extinguisher to 100 psi. using the Schrader Valve.
- Ensure needle on pressure gauge is in the green or safe operating range on the dial.

Stored Pressure Dry Chemical/ 20A -120BC

Use:

- Class A B C fires.
- Mild steel shell.
- Filled with 20 lbs. of dry-chem and pressurized to 195 psi. @ 70 degrees F.
- Total weight 32-33 lbs.
- Has a range of 5'-20' for 8-25 seconds.

Testing:

- Hydrotest every 12 years.

After use:

- Send to storekeeper for replacement.

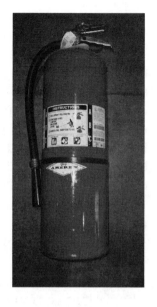

Carbon Dioxide/ 10 BC

Use:

- Class B C fires.
- D.O.T. aluminum pressure cylinder.
- Filled with 15 lbs. of CO_2 and pressurized to 850 psi. @70 degrees F.
- Shell tested to 3000 psi.
- Has a range of 6'-8' for 15-30 seconds.
- Recharge if 1/3 of weight has been lost.
- Total weight is stamped on handle assembly.

Testing:

- Hydrotest every five years.

After use:

- Send to storekeeper for replacement.

FIRE NOZZLES

INTRODUCTION

Fire nozzles are used to deliver water from the hose or pump to the fire itself. For greatest effectiveness, the nozzle must deliver the water with adequate velocity to reach the fire and to reach it in the most effective pattern. Historically, the first nozzles were of great length with a gradually diminishing area within so as to reduce turbulence and increase velocity. In the 1940's, the need for a spray stream was recognized and a combination spray/straight stream nozzle was developed. This nozzle was designed with two openings, one with a straight stream, and the other with a spray diffuser.

NOZZLE TYPES

Smooth Bore – Fixed orifice, non-adjustable, which delivers water in a solid stream only. The quantity and velocity of water is determined by the size of the tip. A solid stream does not lose its continuity until it reaches the point where it loses its forward velocity and falls (*breakover*) in a spray. A solid stream is stiff enough to maintain its original shape and attain required reach/height even in a light breeze.

Periphery – A single orifice nozzle that can provide either a straight stream or fog pattern. One of the shortcomings of the periphery nozzle is that the flow varies as the stream is changed from fog to straight stream because of reducing the orifice size. In other words, the stream is "pinched" down by the baffle when in a straight stream pattern, reducing the flow through the nozzle, as compared to a wide fog pattern. An example of this type of nozzle is a garden nozzle.

Constant Gallonage – Provides a constant flow through the nozzle at various pattern positions. The constant flow or constant gallonage feature provides a means of "shaping" the stream rather than "pinching" it when changing from a fog to straight stream pattern. The barrel or sleeve of the nozzle moves but the baffle remains fixed so the orifice size and flow rate remain constant. To help provide effective streams at the various flows available, ***Select Gallonage*** provides a means of manually altering the orifice size. The manual adjustment moves the baffle in or out, which changes the orifice size and the flow rate.

Automatic – Similar to the constant gallonage nozzle in that it continues to flow the same discharge regardless of the pattern setting. It is designed to automatically provide the correct orifice opening for the flow being delivered to provide a good stream. The basic theory of the nozzle is a pressure sensitive device located in the nozzle assembly that constantly gauges the flow and automatically passes more or less water in accordance with what is available to it. With the nozzle pressure remaining constant, the velocity of the water in the nozzle also remains constant, giving the stream effective reach and pattern.

CARE AND MAINTENANCE

Fire nozzles, like all pieces of mechanical equipment, require periodic cleaning, inspection and lubrication. Inattention to these items may result in difficult or unsafe nozzle operation.

Care – Cleaning and inspection should be done weekly and monthly, depending on usage and storage. Examine nozzle tips for burrs or nicks. Stream pattern adjustment mechanisms and shut-offs should be checked for proper operation, both with and without water pressure applied. Nozzle threads should be checked for nicks and ease of attachment to hose. A good hose gasket should also be used; a protruding hose gasket will deflect the water within the nozzle, causing a poor stream.

Maintenance – Proper lubrication will assist ease of operation. The recommended lubricant is "BREAK-FREE CLP." WD-40 should never be used due to its damaging effect on rubber "O" rings and seals. BREAK-FREE CLP contains high amounts of Teflon, with a minimum of solvents. Lubrication should only be done when needed; manufacturers recommend against excessive lubrication.

Use – Nozzles must be opened and closed slowly. When a nozzle is closed quickly, pressure within the hose line and water main increases approximately four times. This sudden increase in pressure is called "water hammer." Quick opening or closing of nozzles or valves not only endangers the lives of firefighters, but may result in broken hose lines or water mains.

NOZZLES ON E-13

1. Elkhart Brass Select-O-Matic

SM-3FG – Is used on the 1" Booster Lines. These nozzles weigh 4.5 lbs. and have a flow range of 10-75 gpm. As an automatic nozzle, it maintains efficient flow and an effective stream at any pressure. It is equipped with a pistol grip and replaceable spinning teeth that assist in fog stream formation.

TSM-20FG – Is used on the 1 ¾" Hot Lines. These nozzles weigh 5.5 lbs. and have a flow rate of 60-200 gpm. As an automatic nozzle, it maintains efficient flow and an effective stream at any pressure. It is equipped with a pistol grip and replaceable spinning teeth that assist in fog stream formation. These nozzles feature "One for One Hydraulics," which means for every 1-psi increase of discharge pressure it will increase the flow 1 gpm.

STSM 30-BPA – Is used on the 2 ½" handlines. These nozzles weigh 11.5 lbs and have a flow range of 75-325 gpm. As an automatic nozzle, it maintains efficient flow and an effective stream at any pressure. It is equipped with solid "D" vinyl coated handles and replaceable spinning teeth that assist in fog stream formation. A version of this nozzle but with stacked smooth bore tips is also on the engine. The tip sizes include 1", 11/8" and 1 ¼"; the gpm rates for these tips are 200, 250, and 325 respectively.

KK Thunder Fog FT200 – Is carried on all high-rise hose packs. It is a select/constant gallonage nozzle with the following flow rates: off, 30, 60, 95, 125, 150, 180, 200 gpm and a flush setting. This nozzle has a separate bale shutoff that can be used independently of the nozzle to shut down a hoseline. This then permits the nozzle tip to be detached to extend the hoseline.

2. Elkhart Whirling Distributor – This nozzle should have a shut-off near the distributor to control the flow of water. The distributor has a 2 ½" female fitting with four nozzles set in a revolving head.

Elkhart Aeration Tube 246S & 247S – These tubes attach to the 1 ½" and 2 ½" Elkhart Select-O-Matic nozzles and weigh 4 lbs. each. They are used for adding more air to foam solution in order to achieve greater expansion rates. They feature a cast finish with nylon cord wrap.

3. Akron Master Stream Nozzle 5060 – This is an automatic nozzle that has the constant gallonage feature. The flow rate for this nozzle is 250-1250 gpm.

Akron Master Stream Tips with Stream Shaper – A selection of tips are available for use on the Apollo Monitor: 1 3/8", 1 ½", 1 ¾", and 2". The flow rates for these tips are 500, 600, 800, and 1000 gpm respectively. The Stream Shaper is designed with built-in fins to maximize reach.

THE PARTNER SAW

INTRODUCTION

The Partner Saw has become a vital tool in today's fire service. Its primary function is forcible entry, but with a full complement of blades, a variety of materials may be cut. The Partner Saw was introduced into the fire service in 1958, when it was used for crash rescue. Since that time the fire service has taken the saw and applied it to many different operations.

K650 Active-Partner Saw

Location

Various models of Partner Saws are located throughout the LBFD on

- Truck companies
- TRV
- Airport 3
- Various squad, engine and rescue companies.

Safety

- When cutting masonry, water should be applied to the blade and material before and throughout the cut.
- If the blade is forced into cuts, fragmentation can occur.
- Keep cuts straight; binding can cause damage or fragmentation can occur.
- Carbon monoxide can occur when used in confined spaces.
- All personnel within 100 feet must wear full safety gear including goggles.
- Metal cutting sparks create a hazard of igniting flammable materials.
- Gyroscopic effect may affect the ability of the operator to control the saw.
- The weight of the saw can cause fatigue and loss of control.

Specifications

Engine

- Two stroke, air cooled
- 4.8 hp
- 71 cc displacement
- Centrifugal clutch engages at 3100 rpm

Fuel

- 50:1 pre-mix fuel (1 gallon fuel to 2.5 oz. of 2-stroke oil)
- Tank capacity – 26 oz.
- Full tank cutting time – approx. 40 min.

Carburetor

- Tillotson HS – 175H Smart Carb

Ignition

- Magneto – generates 20,000 volts
- Spark plug – NGK BPMR 7A, with a .020 inch gap

Cutting specs

- 12" blade has a 4" depth of cut
- Blade speed – 5500 rpm or 212 mph

Special Features

- 3 stage Active Air Filtration
- Adjustable blade guard
- Decompression valve reduces force necessary to start the engine by 50%
- Fully enclosed Dura-Start prevents dust from reaching starter's essential components
- Manual belt tensioner

- Reversible cutting arm allows for two positions:
  - In-board; gives the saw operator more stability for control during cutting
  - Out-board; allows the operator to get closer to object being cut, or more height during horizontal cuts

Care and Maintenance

- Start saw daily.
- Refuel and clean after every use.
- Change blade after cutting operations.
- Clean any "slag" out of blade guard.
- Inspect blade for proper mounting and saw for any abnormal wear.
- Periodically check foam and paper filter:
  1. Foam – Wash with mild soap and water, dry and re-oil.
  2. Paper – Gently tap out any debris.
- Replace drive belt as needed. A new belt should be run for two minutes and re-tensioned.

Active Air Filtration

The "Active Air Filtration" is a filter system which effectively cleans the air entering the engine in three separate stages, utilizing three different cleaning principles.

- First stage – Separates the cutting dust from the intake air through centrifugal force.
- Second stage – Remaining dust gets stuck in a foam filter saturated with oil and has a large surface area.
- Third stage – Made of pleated paper; provides additional reassurance that the intake air is free from dust.

With "Active Air Filtration," up to 90% of all the cutting dust is separated already in the maintenance free centrifugal stage.

Smart Carb

An internally compensated carburetor with an air duct that links the fuel chamber and filter chamber. This ensures that the air pressure in the fuel chamber and filter chamber remains constant at all times. This design results in:

- High and more uniform engine power
- Better filter economy, longer service intervals
- Lower fuel consumption
- Lower emissions

Blades

1. Metal (aluminum oxide)
   - Always carried on saw.
   - Used on all types of metals.
   - No wet cutting.
   - Blade deteriorates as it cuts.
   - Designation: H12HD (H=1"arbor, 12=12" blade, HD= High Density)
2. Masonry (silicone carbide)
   - Used for brick, block, and soft metals such as cast iron, aluminum, copper, and brass.
   - Wet cut (before and during) to reduce dust, prolong life of blade and reduce chance of thermal shock.
   - Blade deteriorates as it cuts.
   - Designation: H12B (Same as above only the B is for Block.)
3. Carbide tip
   - Cuts wood, lexan and plastic.
   - Tips must point forward.
   - 12 tips per blade.

4. Diamond
   - Used for wet cutting of concrete
   - 16" diameter and only carried on the TRV

Applications

The Partner Saw has many cutting applications:
- Forcible entry
- Auto extrication
- Masonry
- Cable
- Ventilation operations
- Concrete
- Aircraft rescue
- Confined space rescue

Starting Procedures

1. Shake saw to mix fuel.
2. Place red stop switch in RUN position (IN).
3. Cold start: Place choke in START position (OUT).
4. Warm start: Place choke in RUN position (IN).
5. Lock throttle in the open position.
6. Step into footplate or drop start.
7. Pull starter until saw coughs.
8. Move choke to the RUN position.
9. Continue pulling starter until the engine starts.
10. Immediately release the throttle once the saw is running.
11. At idle the blade should not turn.
12. Enter and exit cuts at full throttle.

13. Push red STOP switch down to stop saw.

14. Replace the blade.

15. Refuel and return to apparatus.

Blade Replacement

1. Replace blade after every use. Mark it "training" and store for later.

2. Loosen blade flange bolt.

3. Remove blade and clean blade guard.

4. Metal and masonry blades should be inspected for cracks or damage before use.

5. Center new blade on the arbor and flush with the inside flange.

6. Install the outside blade flange.

7. Tighten the blade flange bolt.

8. Blade should spin easily within blade guard.

9. New blades should be run at full throttle for 30 seconds to ensure proper alignment.

Troubleshooting

Starting Problems

- Check starting control to ensure correct position.
- Check fuel level.
- Check spark plug. If fouled, remove and replace.

Cutting Problems

- If blade speed decreases or bogs, ease off pressure exerted on object being cut (feed speed).
- Check belt tension.
- If abnormal vibration, check blade for damage.

Belt Problems

- If wearing quickly, check drive for abrasions or damage.
- If burning or slipping, adjust tension.

Notes:

Failure of lightweight trusses is one of the many hazards facing firefighters.

Dangers of the Job

Being a firefighter is an extremely dangerous profession. According to statistics, firefighting accounts for the highest percentage of work-related deaths in the United States. It ranks with coal mining as one of the most dangerous occupations in America.

Why is the risk so great? Firefighters are the only people who run into a burning building while everyone else is running out. If there is a chance that a person is trapped inside, we will risk our lives for someone we have never met. It is a code of ethics that we all subscribe to.

Are there things that you can do to minimize the chances of becoming a statistic? Absolutely! The first thing that comes to mind is to ensure that you and your crew are trained to the highest level possible. Knowing your equipment is of paramount importance. It is imperative that a firefighter knows how his or her equipment functions, literally, with eyes closed.

Firefighting is one of the most dangerous occupations.

In extreme situations you will not be able to see your hand in front of your face and it will be so hot your helmet will melt on top of your head. As long as you stay low (on your belly), you will survive. If you stand upright you will literally burn to death. It is crucial to stay low and keep moving. Your number one objective while dragging a hose line through a structure is to locate trapped victims. If it is hot for a firefighter in full protective clothing, imagine how it would be if you were trapped in the environment without protective clothing.

The window in which to save a citizen is very short in duration. As a firefighter you are fighting against time. While you are trying to locate trapped victims, the truck company is on the roof above preparing to open it for vertical ventilation. Since heat and smoke rise, the higher atmospheres are in the explosive range. Once the roof is opened and the heat and smoke are released, the firefighters inside the structure have a better chance to advance the hose line to search for victims and locate the seat of the fire. The relief once the roof is opened is almost immediate. Interior firefighters will feel the temperature decrease as visibility increases. If it's cooler for the firefighters, it is also cooler for the trapped occupants.

As an interior firefighter, you are concerned with the roof failing and falling on you. The newer lightweight construction found in most buildings built since the mid 1970's are a great danger to firefighters. These buildings are constructed with a combination of trusses, which look great to a building engineer on paper, but do not stand up well when subjected to fire. When a building is built with large pieces of wood, it will stay upright even if one or two of the main supports fail, as the other structural members will support the load. In addition, since the lumber has a lot of mass, it will be able to absorb more heat and burn longer before giving way.

Truss construction, on the other hand, is designed to stay upright by a combination of members built in a triangle that collectively give strength to the truss. These trusses are usually held together at the corners by metal pins or surface connectors. The connectors are by far the weakest part of the support system. Since the roof is engineered to take the load, if one of the trusses fails, the entire roof will collapse. Engineers do not factor the failure of one truss member into their calculations. In other words, the entire roof will fail if one small connector gives way when subjected to the heat encountered in a fire.

Unfortunately the dangers of the above scenario are repeated hundreds of times each day across the country. If the interior crews waited for the roof to be opened prior to entering the structure, their job would be much safer. The trade off, unfortunately, is that you would never save anyone who was trapped within the structure. Interior firefighting and vertical ventilation must be performed as a synchronized event.

Performing vertical ventilation is also a very dangerous operation. Imagine for a moment that you are a roofer. The worker's compensation premiums for a roofing contractor are through the roof (no pun intended). This is because being a roofer is an extremely dangerous profession. Imagine walking around on the roof of a building during the daytime looking at how the roof is built. Determine how you would cut a hole in the roof without jeopardizing the structural integrity. One cut of a vital member could result in bringing the roof down on you and anyone inside the building, which would result in certain death for you and anyone below you.

Now imagine walking on the same building at night. Try to locate the structural members. It is a much more difficult proposition. Now put a fire below you and add smoke for good measure. Lastly, imagine hearing the firefighters below asking for vertical ventilation to reduce the heat.

The more familiar you are with the buildings in your district, the better you will perform during the heat of the battle. Preplanning is a critical part of firefighting. It is a matter of stacking the deck in your favor.

Although fighting fires is one of the major dangers we face as firefighters, it is by no means the only danger encountered. Just getting to the calls is a dangerous proposition. We are racing through the city, running through red lights (clearing the intersection before entering), because permanent irreversible brain damage to a non-breathing victim occurs in as little as 4-6 minutes. Time is critical.

Another significant danger firefighters face is dealing with hazardous materials. Some of the chemicals that travel on our roadways can kill a human being with no more than a drop of product on the bare skin. Our structural turnouts are not effective against most of these potent chemicals. In most communities, the fire department is the lead agency in dealing with hazardous materials.

The fire department is also the first to respond to emergencies dealing with compressed gasses such as propane and tankers full of gasoline and diesel. Ignition of these can level a city block.

Terrorism is another significant danger firefighters face. In years past, firefighters never considered themselves as targets. Now terrorists have developed secondary devices designed to detonate five to ten minutes after the initial blast. They calculate that the firefighters will be in the middle of their operations. Their intent is to change the way we do business and instill fear in the community. If they can eliminate the firefighters, who will mitigate the problem? Firefighters in Atlanta, Georgia have twice been the target of secondary devices.

One of the most unexpected dangers comes from those who are elected to protect us, our politicians. The decision to take a firefighter off the engine or truck company, or close a station altogether, is a direct hazard to our safety as well as to the citizens we are sworn to protect. In order to save money, the politicians are willing to gamble with our lives. Often the public is completely unaware of the reductions in public safety personnel.

Being a firefighter is a dangerous job. We all know there is a chance that we will not go home to our families in the morning. Being a firefighter is not a job or a career, it's a calling!

Other Occupations

Should I Become a Paramedic?

Becoming a paramedic is a great challenge.

Instead of putting all of your efforts into getting into paramedic school, I recommend that you first get your education in fire science courses. There are always plenty of firefighter openings that do not require a paramedic license. Getting your education and concentrating on becoming a firefighter, rather than a firefighter/paramedic, will much better serve you.

Many fire departments that are hiring "paramedics" also require that candidates have completed a basic fire academy prior to filling out an application. Even if the candidate has a paramedic license, he or she is often not eligible for hire if he or she has not completed a basic fire academy.

If a candidate does manage to land a job as a firefighter/paramedic prior to completing a basic fire academy, he or she is still going to be held to the firefighter standard. The candidate's chances of completing the academy are not that good if he or she is touching hose and ladders for the first time. Most major departments have a 25-40% attrition rate in their entry-level academy. In other words, if a candidate has not completed a basic fire academy, his or her likelihood of completing the fire department's academy are greatly reduced.

In order to send a firefighter for paramedic training, the department has to cover his or her assignment during six months of school. This is usually accomplished by paying overtime to another willing firefighter. Multiply an average of 10 shifts per month times six months and the department has paid

time and one half (the overtime rate) for 60 shifts. This is above and beyond the $6-$10,000 tuition for the schooling. Of course, there is no guarantee that the firefighter will successfully complete the rigors of paramedic training.

Lastly, with the increasing demand for paramedics, there is no certainty that a fire department will be able to get a firefighter into an impacted paramedic program. It is frequently the paramedics who promote up the ranks, so many departments are fighting a constant battle to keep enough certified paramedics on the roster.

It is true that fire departments want to hire candidates who possess a paramedic license, as it saves departments thousands of dollars. However, I believe that many of the students who enter paramedic school have no idea what they are getting into. They have heard that once they complete training, departments will line up to hire them. This is not entirely true. Although they will be able to compete with a more manageable number of applicants, there is certainly no guarantee of a job offer. Fire departments will not compromise their standards just to hire a candidate with a paramedic license.

Most paramedic schools require 6-12 months of field time as an EMT as a prerequisite to paramedic school. Many EMT's get their field experience by transporting elderly patients to and from convalescent hospitals. While this is an admirable service they are providing, this is not preparing them for the rigors of a paramedic internship. It is extremely difficult for a paramedic trainee to be on top of his or her game when working on a shooting victim, extrication, or patient with severe shortness of breath or chest pain, when his or her prior EMT experience has been limited to being an IV pole (holding the IV bag), or transporting elderly patients to and from their doctors appointments.

The first 12 weeks of paramedic school are called the didactic phase. Classes run from 0800-1700 daily. The days are packed full of in-depth information. Experts compare the plethora of information given during the first 12 weeks of paramedic school to the first year of medical school. Each day different lecturers cover a variety of topics.

Each evening the student has three to four hours of studying to prepare for the next morning's exam. It is required that a student maintains a score of at least 75% to remain in the program. At the end of each week there is a block exam that covers all of the previous week's information. Students must score at least 80% on the block exams. Removal from the program occurs when a student fails two of the daily quizzes or scores less than 80% on a block exam.

Emergency scene simulations are used to help students learn how to identify the patient's chief complaint and determine the proper treatment. One student will exhibit common signs and symptoms of an ailment while the other, similar to a detective, will try to determine what is medically wrong with the patient.

On weekends and weeknights paramedic students are expected to memorize their drug cards. A paramedic must know everything relating to the roughly 25 drugs that are carried on the paramedic unit. Since paramedics administer these drugs, it is expected that all students are able to recite each drug classification, indications for its use, contraindications (when not to use), expected effects, negative side effects, dosage for adults and children and the proper administration rate. Just memorizing numerous facts about each drug is enough work in itself. Once you add it to the daily rigors of the overwhelming didactic responsibilities, you have your hands full.

The next four weeks are clinical rotations where a student does rotating shifts in the local emergency room. Paramedic students complete patient assessments on literally hundreds of patients, become proficient starting IV's, perform advanced airway maneuvers (intubate) and administer the drugs that are carried on a paramedic unit.

During the next two months each paramedic trainee is assigned to a two-person paramedic team. The trainee is expected to perform as a competent paramedic by the 20th shift. The trainee must orchestrate each call, determining the patient's chief complaint, selecting the proper drugs for the situation and basically functioning independently.

It can be overwhelming for a paramedic trainee to run a call. It is difficult to give directions to the firefighters on scene and remember the medication and transportation protocols when seeing that type of call for the first time. If the paramedic trainee does not step up and take control, the firefighters will intervene and take over the run. What makes it even more difficult is the fact that on the majority of fire departments, all of the engineers and captains are

former paramedics who have promoted to the next position.

Each call is critiqued and reviewed with the paramedic preceptors, and improvements will be suggested. It is not easy for a trainee to recover after a bad call when he or she is running 20 calls per shift.

My biggest concern is that candidates did not obtain adequate experience prior to getting into paramedic school. If they have never seen the way the call is supposed to flow, there is no way they can orchestrate all of the things that need to be done. The old adage of "see one, do one, show one" has been fractured. There is no way they can "do one" if they never had a chance to see it done correctly!

During the course of the paramedic internship, the trainee will also be responsible for providing daily drills to the paramedic preceptors and firefighters on the crew. Drills may range from EKG interpretation to cardiac drug therapy. During the shift the trainee is barraged with job-related questions. He or she is expected to demonstrate 100% proficiency, since being correct 70% of the time means he or she does the wrong thing 30% of the time.

I feel that the reason there is such a high attrition rate for paramedics is because they are led to believe the training is not that difficult. I assure you, the training is incredibly difficult. There is a great chance that a candidate will not make it.

Even if a candidate has completed paramedic training, he or she may not have been through a basic fire academy. Being a paramedic does not help you when it comes to pulling hose and throwing ladders.

Most major departments are going to put you through their own fire academy, regardless of your prior experience. A department sponsored fire academy is certain to be much faster paced than the junior college fire academy. If you have completed a basic fire academy, you will have a much greater chance of making it through the department's academy. Depending on the department, there can be a 25-40% attrition rate.

My advice to candidates is to get a fire science education. Work toward an Associate's degree in fire science, since these classes are prerequisites. The next phase would be to put yourself through a basic fire academy in which you will learn how to throw ladders, pull hose, fight fire and all other aspects of being a firefighter.

Upon graduation, become a reserve firefighter so you can get experience, train and maintain your skills. Ideally, the department will be an active one where you are able to see some fire, be involved in vehicle extrications and perform patient assessments. If not, it is imperative that you train on the skills you have learned in the academy. There is no substitute for having your hands on the equipment. It is impossible to learn and maintain the mechanical skills of the job while reading a book in the fire station library. Pull the equipment out and train with it. It is also a great opportunity to learn from the experienced firefighters.

The longer it has been since your academy graduation, the more you forget your skills. It is imperative that you do something to maintain them. Choose a job in the construction industry so you can see how buildings, apartments, or houses are put together, as well as learn the operation of basic hand and power tools.

Most importantly, learn about the hiring process. Understand what the fire department is looking for in the oral interview process. Since most fire departments make the physical ability and the written tests pass or fail, the oral interview is weighted 100% of the overall score. If a candidate focuses on how to take an interview, he or she will be miles ahead of the competition.

If you are still having a difficult time getting hired, you may want to consider getting experience on an ambulance. I would not waste my time doing interfacility transports (IFT's), as you are not learning about how things work in the field. If you are not running first-in 911 calls, you are not getting the maximum exposure to emergency scenes.

As you are running medical calls as an EMT on an ambulance, introduce yourself to the paramedics and firefighters. Let them know that your goal is to become a firefighter (and ultimately a paramedic). Ask pertinent and intelligent questions. Most importantly, keep your eyes and ears open to the rigors of the job.

If you have been running first-in 911 calls for a year and have not yet had a fire department job offer and you have worked seriously on your interview skills, I would only then consider going to paramedic school. Becoming a paramedic will greatly enhance your chances of getting hired. This fact is irrefutable.

It is important to realize, however, there is a wealth of information to learn to become a proficient firefighter. A rookie firefighter must be proficient with all of the tools and equipment carried on the engine and must be a competent EMT. He or she must demonstrate competency in ventilation, ladders, hose lays, salvage and overhaul, as well as fire department policies and procedures. As you can see, there is a tremendous amount expected from a rookie firefighter.

A rookie firefighter/paramedic has all of the aforementioned responsibilities PLUS must be proficient with the paramedic aspect of the job. The crew will not ignore the firefighter responsibilities; since the rookie is already a paramedic, they will expect twice as much from the rookie firefighter/paramedic.

It has been my experience that candidates who try unsuccessfully to get hired in the fire service never learn to take an interview. Instead of learning the basics, the candidates try to get more qualifications than the next person. First they try to take more fire science classes than the next candidate; next it's the fire academy and becoming a reserve firefighter. During this time the average candidate is involved in the fire department testing process.

Most candidates will usually score poorly because they have not learned much about the fire service. More importantly, they have not learned how to take a fire department interview. Their interpretation of why they are not getting hired is that they did not have enough related classes. As they continue to gain more classes and related experience, their interviews usually improve slightly, but not significantly and certainly not enough to land a job.

Their oral interview skills never get any better because they are answering the questions incorrectly. If they had learned how to take an interview, they would have been offered a job months ago. It's not about having more certifications and qualifications than the next candidate; it's about being the person we want to have on our crew.

Think about your friend or classmate who has landed a job. He or she has not had nearly the education, experience or training as many of the more "qualified" candidates. Most just assume the candidate was "lucky," rationalize that the department was looking for women or minorities, or figure that the individual must have known someone on the department. Whatever the rationale, the candidate rarely takes an introspective look at himself or herself.

The candidate continues to stack his or her resume with qualifications instead of learning about the interview process.

Now consider the candidate who is filling his resume with fire-related stuff. He finally gets the word that if he adds a paramedic certificate to his arsenal he is assured a job. So he follows the course and after six months and $30,000, he is a paramedic. He then takes an interview for XYZ fire department that requires a fire academy and a paramedic certificate.

The written and physical ability tests are pass or fail, while the interview is weighted 100%. We know that our case study is a poor interviewer because he could not get a job as a firefighter (that's why he went to paramedic school). Fortunately, all of the other applicants are also poor interviewers (because they too had to go to paramedic school). The department is in a bind; they have to hire recruits, but since they required a paramedic certificate and the fire academy, the applicant pool is small. So, you guessed it, our case study is offered a job.

The message you hear through the testing community is, "I tested for five years, a total of 75 tests, and never had a job offer. I completed paramedic school and got a job within six months." The message he or she sends is loud and clear: Beg, borrow or steal to become a paramedic. It's the only way to get a job.

This candidate now has a job and believes he or she is qualified to give advice to aspiring candidates who attentively listen because the message is coming from a firefighter!

My take is a little different. Because this candidate never learned to take an interview, he or she had to resort to going to paramedic school to minimize the competition. If the candidate had initially worked on interview skills, he or she would have had a job a long time ago!

Notes:

Police officers and firefighters are frequently cut from the same cloth.

Becoming A Police Officer

There is a large contingent of candidates who believe that if they become a police officer first, they will have a better chance of becoming a firefighter later. This does have some degree of truth to it, but I would strongly advise against this route.

If you become a police officer but are really not committed to the job, you can be a hazard to the community and to your partners. Most of all, you are not being truthful to yourself.

There is a great concern that you will hurt yourself, or worse yet, your partner because you are not focused in your job. Any time you are simply going through the motions, you are a liability. There are numerous circumstances encountered every day that require a police officer's good judgment, quick thinking and dedication.

If you are not committed to your goal of becoming a police officer, there is a good chance that the drill instructors are going to key into it and you will not make it through the initial training academy. Your failure to complete the police officer training could jeopardize your chance of being hired as a firefighter.

Imagine how your fire department interviews will flow while trying to explain why you failed the police academy. I can see the oral board members scratching their heads: "Let me get this straight. You tried to go through the police academy but failed. You really didn't want to be a cop, but you thought it would help you become a firefighter?" It's important to note that you are being graded on whether or not you exercise good judgment. This logic will appear flawed to most people.

Take all the time and energy that you would have expended getting into the police academy and focus it on achieving your real goal. It will take at least six months to complete the police academy. Add a year of probation as a police officer and now you are out of the fire department testing circuit for at least 18 months. You have put all of this effort into something you didn't really want. When you really think about it, it doesn't make much sense. You must also consider that any fire department training you have already received has become "stale" because it has been at least 18 months since you received the training.

Will being a police officer help you get a job on the fire department? Absolutely! A police officer has a tremendous amount of responsibility. He or she rides around in the community and makes critical decisions on the spot. If an individual is able to demonstrate this level of commitment and presence of mind, imagine how he or she would be able to contribute when coupled with two or three other crewmembers.

Another reason police officers do well on the fire department is that the screening process to become a cop is very intensive. If someone has made it through the background check, the medical exam, the physical ability and psychological exam, he or she will probably do well in the fire department process. Similar to the police department process, the fire department separates out the less qualified candidates in the testing process, the academy and the probationary period.

My department has hired several police officers, almost one per recruit academy of 24. As a general rule, each of them has done extremely well. They are composed and capable individuals. They fare well in the academy and work extremely well under pressure. They all feel very fortunate to have been hired by the fire department. As a bonus, they were all in the same retirement plan, so they did not lose anything by making the transition from police to fire department. It was a win, win situation for the city and the employee.

There are potentially some negative considerations in hiring police officers. First of all, any time someone who is used to working in an unsupervised role is put in a team environment, there is an adjustment. For some it is very difficult to go from being in charge and making decisions independently, to riding backwards on a fire engine and having every move scrutinized while on probation. Some acclimate well to the teamwork concept, while others find the transition difficult. It is an individual thing that cannot be predicted, as each person is different.

Police officers deal with a tough side of the community. While the vast majority keep a positive attitude, some develop a negative side. This is true with any cross section of the population, but it is magnified when one is a police officer. People often equate the roles of police officers and firefighters, but the jobs are worlds apart. People welcome firefighters. The general sentiment is that everything will be OK now that the firefighters have arrived. On the other hand, the general public does not hold police officers in such high regard. It is unfortunate, but it is a reality. As a firefighter it is easy to keep a positive attitude. People are happy to see you. As a police officer, people want to know why you didn't get there sooner, or why you are writing them a citation.

Being a police officer is a noble profession. Don't do the profession an injustice by entering it half-heartedly. It will be counterproductive for you and everyone around you.

Notes:

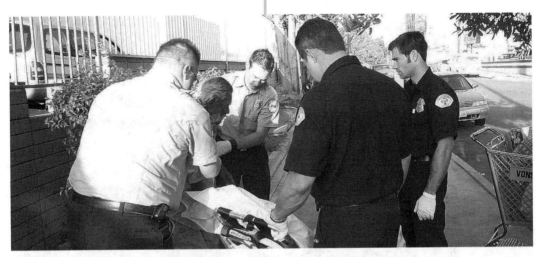

*Working alongside firefighters provides candidates
with excellent practical experience.*

Driving an Ambulance

Driving an ambulance can be a rewarding and educational precursor to becoming a firefighter. If your goal is to get experience, it is important to make sure that you are assigned to a rig that responds to 911 calls. Interfacility transports are important to our society, but to an aspiring firefighter they become routine pretty quickly.

There are many positive aspects of running first-in 911 calls. First of all, the knowledge a candidate will gain working side-by-side with firefighters and paramedics is invaluable. There is no substitute to experiencing the sounds, sights and smells of an emergency scene. You can read about how a scene flows in a book or in the morning paper, but there is no substitute to actually experiencing it first hand.

As an EMT you will have the opportunity to perform patient assessments. Since roughly 90% of the average fire department's calls for service are medical aids, it is critical that a firefighter possess strong EMT skills. The more they are utilized, the better the skills will be.

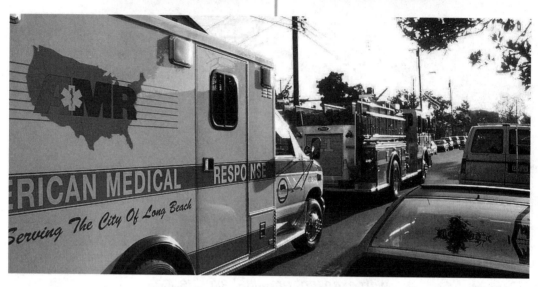

An EMT who is working alongside a firefighter or paramedic and shows a genuine interest in learning and becoming proficient will certainly be shown the ropes. As the professionals who work around you become confident in your abilities, you will find more opportunities to get into the action. In turn, you will be permitted to perform more of the patient assessments. An energetic, friendly EMT who is willing to help out and always has a smile will be welcome on any medical scene.

As an EMT you will have the opportunity to drive Code 3 (lights and siren). Needless to say, the responsibility of driving lights and siren is a mammoth one. It is mandatory that you first pass the ambulance operator exam at the state Department of Motor Vehicles. After you have earned the respect of your peers and the trust of the management, you will be assigned to an ambulance.

An EMT is responsible for maintaining and stocking his or her ambulance. Routine maintenance and inspection are a common part of every shift. At the beginning of each shift, an EMT must go through each compartment of the ambulance to make sure the rig is stocked and ready to go. These same duties are performed on the fire engines and truck companies every shift at the fire station.

An EMT who becomes complacent in his or her job is not maximizing his or her opportunity. It is imperative that each day is used as a new learning experience. On the long mundane days, between calls the diligent EMT will pull out a piece of equipment and reacquaint him or herself with it. It may be a few months between calls in which an EMT is tasked with applying a Hare Traction splint. Even though it has been a period of time, the EMT is still expected to be proficient in the skill. There is no faster way to have your peers (and the patient) lose confidence in your ability than to be stumbling with a piece of equipment on an emergency scene.

Many private ambulance companies have an agreement with the local fire agency to "house" their ambulances in the fire stations. This is a golden opportunity for an aspiring firefighter to not only work side-by-side with the firefighters, but also gain insight into the job and lifestyle desired. You already understand what it takes to get along, you are familiar with station life and you are working in the field. How much of a stretch is it for a fire agency to provide additional "firefighter" training and change the color of your shirt?

Since one of the most difficult parts of being a firefighter is getting along with others and living and working together for extended periods of time, a fire agency knows (or will be able to verify with a single phone call) that you are able to work within the structure of the fire service. If an individual does not fit into the family network, he or she is not welcome to work in the fire station. An EMT who is well-liked by the other firefighters in the station will very likely get hired and become a firefighter in the station. It's important to remember that we hire people we like. We can train you as a firefighter. We can't train you to get along with others.

Since firefighters have a strong network and bond, we know firefighters on surrounding departments. When a well-liked EMT is testing for a neighboring agency, a personal referral from a firefighter who works side-by-side with that EMT is monumental.

Notes:

Different Perspectives

Female Firefighters

Sharon Easley is a firefighter for the Long Beach, California, Fire Department. She has graciously agreed to share her experiences of getting hired and working as a firefighter for a large metropolitan fire department.

My fire career began in January of 2000 when I was hired on with the City of Long Beach and was about to begin the department's 14-week academy.

I was on my way to fulfilling my ultimate career goal of becoming a firefighter. Throughout the academy, I remained determined to let nothing get in my way of accomplishing that goal. It was during this time that I truly realized that my time spent going to school, volunteering, and training would all boil down to this academy. This is where it all comes together. And sure enough, it came together.

The steps I took to become a firefighter were similar to those of many other firefighters who worked hard to get into the fire service. I enrolled into a community college Fire Science program. While completing the required classes, I continued my training program in the gym.

My training routine became focused around job specific techniques and exercises. I concentrated heavily on upper body strength. Generally speaking, that is where women have problems. Face it, women aren't built like men; they can't muscle everything. The use of proper technique comes into play in order to get the job done. If this is understood, then hardcore training on upper body strength should be a priority.

While keeping up with training, I also signed up for physical ability practice classes that helped improve my skill level for future agilities. I knew that I had to work overtime on the physical aspect of this endeavor if I wanted to succeed. This routine really helped me pass the agilities required by hiring departments. After completing all of the fire science classes, I passed the ability test that assured my eligibility for the upcoming part-time college fire academy. After my name was chosen from a lottery, I started the fire academy and graduated nine months later, accomplishing the first milestone on the path towards my goal.

After the college academy, I took a job working as an EMT for a private ambulance company so I could gain more experience in the field. I then applied for the Santa Ana Fire Department Reserve program. I was hired on and volunteered there for a year and one half. Being a reserve for Santa Ana was an awesome experience. I was able to work as the fourth person on the engine and fifth person on the truck. I experienced some great medical and fire calls and learned about station life as a rookie. The captains worked with all of the reserves on oral interview skills and attaining a Firefighter 1 certificate. My time with Santa Ana really prepared me for my career with Long Beach.

As a female in this career, I knew that entering a male oriented occupation was not going to be a walk in the park. I believe that I am here to do a job, not to change the fire station or fire service. I didn't want to come into the station and disrupt the dynamics, but be a part of those dynamics. And I have been lucky to be surrounded by some pretty great guys. A female co-worker of mine once told me that if I always kept my head up, worked hard and had a good attitude, that I would make it in this "place." I'll never forget those words. That was probably some of the best advice that I had ever received.

Although I am young in my career, this department has made me feel welcome for many reasons. I guess it is what you make of it, and I came in knowing that I want a long, happy, fun career. I have learned so much from this job: not just how to do it, but how to get along with others.

Growing up with a household of brothers helped, but I still had to "pass the test" within this job. Here is one example: it is only obvious that working 24 hours with a group of men is nothing like working with a group of women. "Table talk" is much different with a group of men. And I believe that if a female co-worker cannot handle that, there are going to be problems. It is all about adapting to your surroundings. I have learned a lot about "a lot" while talking at the table. Many things I am sure I wouldn't have learned elsewhere.

If a career with the fire service is what you want, there are many hurdles to jump over. However, it is achievable...only if you want it

badly enough. And I am sure that most female firefighters will tell you the same. What they might also tell you is how great their job is. I know that I would tell you that. This is the best job in the world. If you are ready to do this, be prepared for a lot of hard work, and talk to other females in the service if you can. Learn about their experiences and use them to your advantage.

Notes:

Different Perspectives

Notes:

_____

_____

_____

_____

_____

_____

_____

_____

_____

_____

_____

_____

_____

_____

_____

_____

_____

_____

_____

_____

_____

_____

_____

Different Perspectives

Firefighting: A Wife's Perspective

By Marian Lepore

From the beginning of our relationship, I knew this would be different. We could only see each other on red and green days and I could only call him at work after 9 a.m. or before 9 p.m. and never at mealtime. No one warned me what it would be like to date a firefighter.

After I met his family, I was introduced to his firefighter family – the three crewmembers he spent ten 24-hour shifts with each month. They knew everything about me. I came to realize that I would have to be willing to share him with his coworkers, both on and off duty.

It didn't take long for me to learn the peculiarities of fire department etiquette. When I visited the fire station for the first time, I had to bring a pie. In fact, whenever a firefighter does something for the first time, whether it's buying a house, being mentioned in the news, or having a child, he or she must bring ice cream for the crew.

On birthdays, firefighters bring in their own cake. When they get promoted or reassigned to a new station, they cook their own farewell meal for their coworkers. It became evident to me that firefighters are more comfortable serving others than being served.

When we became more serious in our relationship and eventually married, the church and reception hall were filled with firefighters and their families. The happiness of one was celebrated joyously with the rest (of course, after all the jokes of bringing running shoes for the groom). The birth of children, purchase of a home, or completion of a college degree is all celebrated as if it were close family members achieving these successes.

I could see that firefighters are bonded in a special way. They spend 24 hours at a time together, which is much more time than most family members spend with each other. They work together for a single purpose, whether it's to save a life, put out the flames in a burning building, or educate children in fire safety. They must be willing to risk their lives for each other without hesitation.

Firefighters take care of each other. If one is going through a divorce, he or she is counseled, supported and encouraged. If another is having difficulties with a rebellious teenager, many others can offer advice from their own experiences as parents. When a firefighter is trying to promote, he or she may carefully choose the next station assignment knowing that a certain crewmember will help with oral interviews or fire simulator problems.

When I first started dating my husband, I couldn't believe that a 23-year-old could own a home. He later explained that when he first started on the fire department, an older firefighter sat down with him and educated him on the importance of saving for and purchasing a home. He also taught him about deferred compensation and how important it is to maximize his contributions from the very beginning. Thanks to the wisdom and caring of this older firefighter and the magic of compounded interest for investments, my husband and I both maximized our retirement savings (his deferred comp, my 401K) and we will retire comfortably.

My husband has carried on this tradition of helping new recruits by educating them on financial investments and deferred compensation. Firefighters look out for each other in every way.

Everything in the fire service is done in a big way. The Long Beach Fire Department has the biggest grill I have ever seen. It is built on wheels and is towed behind a truck. I would have thought it was ridiculous if I hadn't seen that every spot on the grill was being used. This grill is used for graduation ceremonies, department picnics, fund-raisers and all types of community events. Only a firefighter could have dreamed up that grill!

When a firefighter cooks, he or she cooks in a big way. It doesn't seem to matter if it is a large station with several engine companies and rigs, or a station with a single engine company and a crew of four. There are always at least two refrigerators at the station to hold all the leftovers. When my husband is at home, he carries on the tradition and cooks enough to feed an army. I also have two refrigerators in my home.

Maybe firefighters are just trained to think in a big way. But along with big ladders and big trucks come big responsibilities.

When I was dating my future husband, I was a student in the physical therapy program at California State University, Long Beach. I was taking anatomy and physiology classes and was interested in the medical side of his

job. He was still a paramedic at that time and had not yet promoted to captain. He suggested that I ride along with him to see what he did. The television show ER didn't hold a candle to the real life drama I witnessed.

It was pretty slow (he thought) and I accompanied him on calls responding to SOB (shortness of breath) and a drug overdose. We were just sitting down to an elaborate Mexican dinner, when another call came in. It was reported as a man down, gunshots heard. The crew responded immediately to the call.

When the paramedic rig and the fire engine arrived, there was a large, angry crowd gathered. The police had not yet arrived, so it was not known whether the assailant was still present in the crowd or had left. The victim was not even visible through the crowd. The captain, who always looks out for his crew, ensured that the police arrived to control the crowd and clear the area. The victim was a teenage boy with a gunshot wound to the chest.

He was hooked up to an EKG machine, given an IV for fluid and other medications and the bleeding controlled as well as possible in the field. They kept in constant communication with doctors in the ER, so the medical staff could give further instructions and was fully prepared for him when he arrived. Every crewmember was needed, whether it was to take vital signs, control bleeding, administer medication, fetch equipment, use the radio, or interview family members. I was in awe of how efficiently this team could work, with a critical victim in the field, poor lighting, a large, noisy crowd and possibly an assailant who did not want this victim to survive.

The victim was rapidly transported to the ER, where the paramedic team was integrated into the hospital's response and they worked together to try to save this boy's life. Within minutes his chest was cracked open and there was the largest pool of blood I could imagine beneath the gurney. Even with CPR, repeated administration of cardiac medications, defibrillation, IV fluids, intubation and other intensive efforts, they could not save his life. The bullet had nicked his aorta and he had lost too much blood.

His family was in the waiting room. His mother became hysterical and his brother vowed revenge for this gang-related shooting. The crew returned to the station to finish dinner and prepare for the next call. This experience will remain vivid in my memory for the rest of my life. For the crew, it was just another day on the job. They felt compassion for the victim and his family, but they could

The Los Angeles riots put the fire service to the test.

not be overwhelmed by it, or they would not be able to continue working.

Along with the intensity of responding to critical emergencies and the danger of entering burning buildings, there can be unexpected dangers. In 1992, after the verdict in the Rodney King beating was announced, Los Angeles County went crazy. There was rioting throughout the streets. People were burning down buildings, beating total strangers and looting stores. It was out of control.

People were so angry that they were shooting at anyone in authority, including firefighters. As if the job were not dangerous enough! There was one incident that my husband only told me about years later and it was only after a coworker casually referred to it. A call came into the station that a strip mall was on fire. Due to reports of firefighters being shot at and threatened by crowds, they were supposed to wait for the police to show up and accompany them to the scene. The police were busy elsewhere, as you can imagine, so the fire department responded anyway. Just as they were finishing, they were shot at and had to take cover behind the fire truck. They managed to get into the truck safely and quickly left the scene. As they left, they could see the arsonists leaving their hiding place to prepare to burn the buildings again.

Sure enough, shortly after returning to the station they were called out again to the strip mall. This time they put on their flak jackets and waited for the police to accompany them on the call. They put out the fire in what was left of the mall. That was the longest night of my life and I didn't even know how truly bad it was until later.

Spouses of firefighters also support each other. Whether it is by getting together for Bunco monthly, taking care of each other's kids, or just chatting over a cup of coffee, it is important to share any concerns with others who understand. Marriage can be challenging enough for couples who work Monday through Friday from nine to five. Add the stress of dealing with an always changing work schedule, a dangerous environment and the need to be completely self-sufficient, and it can be disastrous for a marriage. The best way to cope is to maintain your sense of humor.

Humor and laughter is an integral part of fire station life. My husband brings home stories of outrageous deeds and unbelievable wit nearly every shift. If late night talk show hosts need new material or writers, they could do no better than some of the creative minds on the fire department. Especially funny stories of practical jokes or extreme composure after being water-dropped become urban legends.

When I was dating my husband and planning to visit him at the fire station for the first time, he warned me to look up before I entered the station. He said that sometimes first-time visitors were water-dropped when they entered the station house. I had no idea what he was talking about. These were adults. He must be joking. Well, I was lucky that my ignorance did not get me into trouble. I remained dry throughout that first visit. It was only later that I realized he was not joking.

I realized immediately that it is not only the firefighters who have to have a good sense of humor. During our wedding ceremony, our exchange of vows was delayed by several minutes as the blaring of a siren just outside the church doors drowned out the minister's words. Later at the reception, one of the layers of cake looked odd to me. When I investigated, I found the inside of the cake had been hollowed out, filled with paper towels and then recovered with frosting. When I turned to my new husband in shock, he just shrugged as if to say, "Of course they cored the cake."

The practical jokes continued at home. Our children learned the hard way that they had to learn to laugh in the face of disaster. Of course, a child's idea of disaster is not exactly the same as an adult's. When our oldest daughter was in elementary school, she worked hard to complete a 'book float,' which is a visual book report built on the top of a shoebox. Her book float was elaborate, with trees made of broccoli tops glued to the shoebox. When she was getting ready to go to school the next morning, she found that all of her 'trees' had been chopped down! Her father had eaten the tops of the broccoli that morning before he left for work. He thought it was a hilarious joke. She did not feel that way. After many tears and an emergency session with a glue gun, she finally began to see the humor in the situation.

Our youngest daughter found that she had to be on guard at all times. One day when she was watching her favorite TV show, she became frantic because the TV kept changing channels all by itself. Her father finally confessed that he was using the master remote control from a distance. Now I find that I am the one who needs to stay on her toes in our house. Our children have learned the hard way to give as well as they got.

Without a sense of humor, a ready joke and the ability to see the bright side of things, the tragedy firefighters encounter every shift would soon overwhelm them. It is a coping mechanism to help deal with the seriousness of the job. If a firefighter candidate cannot laugh easily and often at him or herself, the candidate will either not succeed, or will not be happy on the job. He or she will never understand the culture of the fire service.

After my husband was in a terrible head-on collision between the engine he was on and a police cruiser, he was out of work for several months. He fought to return to work full duty. I think the fire service must be one of the only professions in which its members enjoy the job so much they will not consider an alternative.

My husband shows up at the station 45 minutes before the start of his shift, just in case he can take a call for the captain coming off duty and allow him to leave work on time. When my husband is going off duty, he stays to share a cup of coffee and some laughs with the oncoming crew. I know of no other profession in which its members are not in a hurry to leave after their shift is over.

So why do so many people dream of becoming firefighters?

The fire department schedule is one of the biggest draws to the job. There is no other job in which you can work only ten days a month, with either six or four days off at a time. The problem is that when my husband wants to go on vacation, he doesn't understand that I can't match his schedule and just take off four or six days at a time. At least I know his schedule a year in advance!

Because their schedules are so different from everyone else's, firefighters like to vacation together. It is common to see large groups of firefighter families on vacation in Hawaii, Baja, or Lake Havasu. It's also convenient to share the childcare duties with other parents.

Firefighters generally enjoy their work schedule, but it can be hard on a family and marriage. Spouses must be self-sufficient and prepared to take care of crises on their own. If a firefighter's child is sick, he or she cannot just leave the station to pick up the child from school. It is critical that the firefighter remain at work to keep the station fully staffed for emergencies. If a firefighter goes home, it must be for a serious injury or illness.

The firefighter schedule can also be inconvenient on holidays. Most people are used to spending holidays with their family. Firefighters don't have a choice. If they are scheduled to work on a holiday, they work. Unless they are going out of town, they do not request the day off. Everyone would love to have holidays off to spend with his or her family, but someone must work. If they were to call in sick and no one was signed up to work overtime, another firefighter would be force hired and pulled out of a family gathering. A firefighter spouse must be flexible enough to be prepared to cook and entertain all by him or herself at a moment's notice.

It is expected that the younger firefighters without families offer to work on major holidays. As they get older and have their own families, the favor will be repaid by the next generation of firefighters.

When a firefighter is scheduled to work on a major holiday, the family members are often invited into the station for a holiday meal. The crew will go all out and prepare a lavish feast. Sometimes the family members end up eating all by themselves as the crew is called out on an emergency. The kids don't mind. They feel that the more time in the station, the better. Again, it just goes with the job.

Because they work a set schedule regardless of holidays, firefighters get time off which includes both vacation and holidays. When they take time off, it is usually for several weeks at a time. I have found that having your husband on vacation can be worse than having your kids out of school for the summer. A firefighter with too much time on his hands can get into much more trouble than your kids.

Firefighters are generally do-it-yourselfers. This is why you will often see them in Home Depot. They are mechanically inclined and are used to improvising to solve problems quickly. You may come home from work one day and find that you have a new laundry chute or the washing machine is being rebuilt. If you are someone who likes things done a certain way, then for the sake of marital harmony, I suggest you call a professional out to build, repair, or replace whatever it is before your husband's next vacation, four or six day.

Most of us go on vacation to get away from our jobs. When my husband is on vacation, he seeks out fire stations. I have gotten used to losing my husband for a few hours during a vacation while he rides along with the local fire department. Of course he is hoping to go on a really 'good' call (which to the rest of us means 'bad').

My husband has T-shirts from fire departments in Alaska, Illinois, Louisiana, Nevada, Texas, Washington, Washington, D.C., Utah and many more. I'm almost embarrassed to say that I even visited a fire station on my own when I recently went to New York City. When my husband sees someone wearing a fire department T-shirt, he will always ask that person if he or she is on the job. There is an instant bond between them, to the point that two total strangers can joke and tease each other about their respective departments.

Another advantage about the fire department is the benefits. There are usually many options for medical and dental plans, so you can pick the plan that is right for you whether you are single or married with a family. The credit union can't be beat. They give personalized service and actually know your name when you call.

One of the biggest benefits is the retirement package. It is negotiated as part of the firefighters' contract. Unfortunately, the life expectancy of a firefighter after retirement is not as long as that of a person who has not been

exposed to smoke, chemicals, stress, blood, injury and interrupted sleep for their entire professional career.

Depending on their age when they were hired, firefighters usually retire in their fifties. However, they will not be bored after retirement. Most firefighters have hobbies which take up a great deal of their time, such as skiing, fishing, boating, fishing, traveling, fishing, or biking. When I was dating my future husband, he said he liked to fish. I was thinking that in Southern California, it's only fishing season in the summer. No big deal, maybe I would even go with him sometimes. It was only later that I realized that it is always fishing season somewhere in the world. I should have been forewarned when he had to check his fishing tide book before committing to a wedding date.

Every firefighter has a side business. This business is not reserved for after retirement. They conduct this business throughout their fire service career. Since they work only ten days a month, there is plenty of time off to do carpentry, plumbing, concrete, tile setting, painting, roofing, CPR instruction, writing, or manage whatever business they have invested in. The advantage to other firefighters (and their wives) is that whenever something needs to be fixed at home, there is always a firefighter with the skills to do it. Forget paying full price to a plumber, electrician, or drywaller! By trading skills and services, most firefighters are able to remodel and upgrade their homes.

Firefighters earn a good salary and are rarely ever laid off. Overtime shifts also help immensely. However, I don't know if you can truly compensate someone for the long-term effects of a chemical fire, or the emotional scars from being first on scene at a horrendous child abuse incident. Firefighters seldom talk about the really terrible things they witness, but we all know we can count on them when we're in trouble.

People love firefighters. Children and even some adults wave at them as they drive by on their big trucks. When others accompany a firefighter, even off duty, the benefits often extend to them. After the Southern California wildfires, Disneyland in Anaheim was offering free admission for firefighters and their families as a thank you. We invited our neighbors to go with us. I am a physical therapist, my neighbor is a teacher and her husband is a computer consultant. None of us has ever been admitted to an amusement park for free just because of our profession. Firefighters, however, are universally loved, appreciated and welcomed.

It may seem to outside observers that firefighters all look similar: tall, lean, dark hair and a moustache. Well, departments have changed over the years as they seek greater diversity, skills and strengths. They try to hire firefighters who can relate to and speak the languages of the people in the community. They hire female firefighters who can contribute their abilities and perspective to the department. They even hired my husband despite the fact that he cannot grow a decent moustache.

Maybe one reason that firefighters seem so alike is that they have the same attitudes. They are honest, brave (you wouldn't catch me running *into* a burning building) and exceedingly generous with their time and talents. When they take the time to tutor children, fix up a dilapidated house in the neighborhood, or collect and hand out Christmas gifts to disadvantaged children, it is all on their own time. Their spirit of public service is an example that should humble the rest of us. I can't resent the time my husband takes to help others, because it is part of who he is. Our youngest daughter had a wonderful time one Christmas when she was able to help hand out donated gifts and ride with Santa in his sleigh atop a fire truck.

I have wondered how the fire department manages to hire so many people with the same attitudes. I guess it is because they know what they are looking for. The selflessness and willingness to sacrifice can't be taught. It must be an integral part of their makeup. When a firefighter or family member is seriously ill, others will line up to cover his or her shifts with no expectation of being repaid for their time.

As a spouse, I will never understand my husband's excitement when he is called on to spend days fighting a raging wildfire, or enthusiastically describes in vivid detail the fire that ripped through the chemical warehouse. But his coworkers understand. They will always be there for him, working towards the same goal and watching his back. I count on them to do that.

The fire service is a very large, caring, fun-loving family, of which I am proud to be an extended member. I know that even if my husband is lost at sea during one of his many Baja fishing trips, or something unthinkable happens during one of his calls at work, my children and I will always be taken care of.

Having a Firefighter for a Parent

By Samantha Lepore

There are many pros and cons to having a parent as a firefighter. For example, my Dad can take time off work for a few days at a time, but is at the fire station for a few days at a time, too.

I'm really proud of my Dad for being a firefighter and making a difference in our world by saving people's lives. But it's a dangerous job, and I'm usually worried about my Dad fighting fires.

I really love it when my Dad takes me to the fire station. I remember one time sliding down the pole, over and over! I also got to ride in a fire engine one time.

My Dad once talked to my class about fire safety. My friends loved dressing up in my Dad's turnouts! The boots went up to their knees and the coat touched the ground. They practically got lost in his uniform!

In conclusion, it's really cool to have a Dad as a firefighter!

Did you know...

- That firefighters sometimes rescue hamsters?
- That firefighters usually don't rescue cats stuck in trees?
- That in Michigan it is illegal to chain an alligator to a fire hydrant?
- That if you boil a gallon of water, you would get 1,700 gallons of steam?
- That the first fire hydrant was installed in 1801 in Philadelphia?
- That in 1679 Boston was the first city in America to have a paid fire department?
- That a fire is reported in the US every 15 seconds?
- That about 90% of homes in the US have smoke detectors?
- That about two million fires are reported in the US each year?

Notes:

God's Cake

— Author unknown

Sometimes we wonder, "What did I do to deserve this?"
or, "Why did God have to do this to me?"

A daughter is telling her mother how everything is going wrong, she is failing Algebra, her boyfriend broke up with her and her best friend is moving away.

Meanwhile, her mother is baking a cake and asks her daughter if she would like a snack, and the daughter says, "Absolutely Mom, I love your cake."

"Here, have some cooking oil," her mother offers. "Yuck!" says her daughter.

"How about a couple of raw eggs?"
"Gross, Mom."

"Would you like some flour then? Or maybe baking soda?"
"Mom, those are all yucky!"

To which the mother replies: "Yes, all those things seem bad all by themselves. But when they are put together in the right way, they make a wonderfully delicious cake!"

God works in the same way. Many times we wonder why HE would let us go through such bad and difficult times. But God knows that when HE puts these things all in HIS order, they always work for the good. We just have to trust him and, eventually, they will make something wonderful!

Different
Perspectives

God is crazy about you. He sends flowers every spring and a sunrise every morning.

Whenever you want to talk, HE will listen.
He can live anywhere in the universe, and HE chose your heart.

I hope your day is a "piece of cake!"

Life may not be the party we had hoped for, but while we are here we might as well dance the dance.

Notes:

_____

_____

_____

_____

_____

_____

_____

_____

_____

_____

_____

Appendices

Interview Grading Sheet

City of
RATING STANDARDS
FIREFIGHTER

OUTSTANDING   *(90 - 100)*

Candidate's overall performance and presentation is clear, to the point and thorough; response(s), or skill, reflect advanced knowledge or ability level for the position for which the examination is constructed; clearly exceeds the typical knowledge/ability/skill level of most other competitors and demonstrates nearly all of the ideal characteristics of the rating factor(s). **Candidate would be an immediate asset.**

WELL QUALIFIED   *(80 - 89)*

Candidate's overall performance and presentation is focused on subject and/or issues; response(s), performance, or skill, reflect a mastery of the knowledge or ability measured. **Candidate would function well** with regard to the factor(s) evaluated following a brief period of orientation/training.

ACCEPTABLE   *(70 - 79)*

Candidate's overall performance and presentation is adequate; shows an understanding of the subject area and/or issue, but may not be as thorough as the higher performing candidates; response(s), performance, or skill, reflect the basic knowledge or abilities required to adequately do the job. **Candidate could function successfully** with regard to the factor evaluated, but would need some training and supervision to develop those knowledge, skills and/or abilities, which have not yet been developed.

PASSING SCORE =70

UNACCEPTABLE   *(65)*

Candidate's overall performance and presentation does not demonstrate the knowledge or ability required; does not reflect the competency required to carry out the job. Candidate **is not ready to function adequately with regard to the factor(s) evaluated.**

ORAL INTERVIEW
TEACHING DEMONSTRATION RATING SHEET

CANDIDATE: _____

Rating Factors	1 - 5
Preparation – Got the audience interested in the topic	
Presentation – Imparted new information	
Application – Asked questions and got audience participation	
Summary – Re-capped highlights of lesson	
Instructional Aids – Legible and relevant	
Overall Effectiveness – Kept the audiences interest	
Mannerisms – Presentation did not distract audience	
Communication – Good choice of words, used voice inflections	
Knowledge of subject – Could you handle the fire at the end of the lesson?	
Total Score- 45 points possible	

EVALUATOR: _____ DATE: _____

Lesson Plan Format 5 points possible	

Total Score- from both categories - 50 points possible	

**ENTRY LEVEL
FIREFIGHTER 2000**

CANDIDATES NAME: _____

INTERPERSONAL SKILLS

weak ☐☐☐☐☐☐☐☐☐☐☐☐☐ strong

☐ Relates to the needs, interests of the parties involved
☐ Calm manner
☐ Resolved issues reasonably.
☐ Followed chain of command
☐ Able to sort out facts
☐ Saw the whole picture   ☐ Took control of situation
☐ Gets the facts quickly
Comments: _____
_____
_____

PROBLEM SOLVING

weak ☐☐☐☐☐☐☐☐☐☐☐☐☐ strong

☐ Logical thought process
☐ Figured out all aspects of the problem
☐ Came up with good solutions to the problems
Comments: _____
_____
_____

ETHICS

weak ☐☐☐☐☐☐☐☐☐☐☐☐☐ strong

☐ Knows right from wrong
☐ Handled in manner conforming to dept. policies
☐ Didn't let others sway their thoughts
Comments: _____
_____
_____

TEAM WORK

weak ☐☐☐☐☐☐☐☐☐☐☐☐☐ strong

☐ Got involvement and buy-in from others
☐ Able to get station or crew to help out
☐ Well organized, didn't take on the whole problem by themselves
☐ Team player - involved in some type of group activity
Comments: _____
_____

ENTRY LEVEL
FIREFIGHTER 2000

QUALIFICATIONS

weak ☐☐☐☐☐☐☐☐☐☐☐ strong

☐ Displayed confidence ☐ Education commensurate with job
☐ Broad experience base
☐ Well rounded in more areas than fire and EMS

Comments: _____
_____
_____

OVERALL SCORE

Points	Description
91 - 100	Outstanding candidate. Would enjoy working and living with the candidate on a 24-hour shift schedule. Great potential with no areas of concern. Could hire immediately.
81 - 90	Desirable candidate, but only after top candidates are selected. Would enjoy working and living with the candidate on a 24-hour shift schedule. Good potential
70 or 80	Not a desirable candidate. Has some or several areas of concern. May be a viable candidate at a later date.
Below 70	Do not hire. Would not like to work with this candidate. Several areas of concern.

SCORE _____

Comments: _____
_____
_____
_____
_____

EVALUATOR'S SIGNATURE: _____

Probationary Firefighter Evaluation

> ## FIRE DEPARTMENT
>
> ## PROBATIONARY FIREFIGHTER
>
> ## MONTHLY EVALUATION
>
> | Probationary Firefighter: | | Date: | |
> |---|---|---|---|
> | Station / Platoon: | Captain(s): | Recruit Class: | |

The development of probationary firefighters is one of the most important responsibilities of field Captains. Integral to this development are on-going evaluations. The "Probationary Firefighter Monthly Evaluation" instrument is an important part of this process.

The "Probationary Firefighter Monthly Evaluation" instrument is a Behaviorally Anchored Rating Scale (BARS) designed (a) to assist probationary firefighters obtain accurate self-assessments of their development and (b) to provide the Captain-evaluators with behavioral descriptions for consistently rating probationary firefighters on job-related dimensions. Specifically, the BARS instrument is divided into four dimension clusters:

- Personal Traits
- Station Duties
- Emergency Duties & Training
- Fire Prevention/Public Relation

Each dimension cluster is delineated in more specific sub-dimensions.

The probationary period requires a commitment by both the probationary firefighter and their Captain(s). Probationary/apprentice firefighters must be committed to become competent firefighters and Captain(s) must be committed to managing the probationary firefighters' development. Thus, it is essential for you to understand thoroughly this and other related instruments as they contribute to the development process.

The evaluation will be given in a straightforward manner emphasizing the acceptable as well as the unacceptable aspects of performance and behavior. The Captain(s) must provide specific descriptions of observed and/or demonstrated behaviors on incidents (provide incident number), during drills and interactions with others to justify the given rating.

It is the responsibility of the station captain(s) to evaluate each probationary firefighter on a monthly basis and submit this evaluation form through channels to the Director of Training.

Attach additional pages if necessary.

Please contact the Training Division's staff if you need assistance in the use of the "Probationary Firefighter Monthly Evaluation" instrument.

Probationary Firefighter:		Date:	

PERSONAL TRAITS

Approach Towards Supervisors

☐ Consistently demonstrates an unsatisfactory response toward supervisors. Takes constructive criticism poorly; argues and/or rationalizes mistakes. On occasion does not show proper respect when relating to supervisors or superiors. Lacks enthusiasm and/or initiative in following suggestions on areas needing improvement. Consistently fails to carefully or accurately follow the directions / orders of supervisors or superiors.	☐ On occasion demonstrates an unsatisfactory response toward supervisors. Occasionally takes constructive criticism poorly or with no response. Occasionally argues and/or rationalizes mistakes. On occasion does not show proper respect when relating to supervisors or superiors. May lack enthusiasm and/or initiative in following suggestions on areas needing improvement. Occasionally fails to carefully or accurately follow the directions / orders of supervisors or superiors.	☐ Consistently demonstrates a satisfactory response toward supervisors. Takes constructive criticism well; does not rationalize mistakes. Always demonstrates proper respect toward supervisors or superiors. Demonstrates enthusiasm and initiative in improving areas in need of improvement. Follows carefully or accurately the directions / orders of supervisors or superiors	☐ Consistently demonstrates a satisfactory response toward supervisors and superiors that is above the standard. Always demonstrates proper respect toward supervisors or superiors, even under trying circumstances. Takes constructive criticism well and demonstrates focused, systematic determination in working to correct areas in need of improvement.
Provide specific descriptions:			

Response Towards Other Members

☐ Consistently demonstrates an unsatisfactory response toward other members. Does not take constructive criticism; argues and/or rationalizes mistakes. Lacks enthusiasm in relating to others. Lacks initiative and often fails to perform specifically required tasks in cooperation with others or as one's individual responsibility in groups. Sits back and lets others perform necessary tasks. Consistently fails to function as a team member. (Does not work with others, help others, and/or subordinate his/her goals to those of the group.)	☐ On occasion demonstrates an unsatisfactory response toward other members. Takes constructive criticism poorly; argues and/or rationalizes mistakes. Lacks enthusiasm in relating to others. Lacks initiative and occasionally fails to perform specifically required tasks in cooperation with others or as one's individual responsibility in groups. Sits back and lets others perform necessary tasks, except when prompted. Occasionally fails to function as a team member. (Has difficulty working with others, helping others, and/or subordinating his/her goals to those of the group.)	☐ Consistently demonstrates a satisfactory response toward other members. Takes constructive criticism well; does not rationalize mistakes. Demonstrates enthusiasm when relating to others. When appropriate, takes the initiative with work that needs to be done in cooperation with others or as one's individual responsibility in groups. Works to improve areas needing improvement without being told. Consistently functions as a team member. Learns from other team members	☐ Consistently demonstrates an attitude toward Department members that is above the standard. Demonstrates enthusiasm when relating to others. When appropriate, takes the initiative with work and helps others. Works to identify and correct weaknesses. Demonstrates focused systematic determination in seeking to be a team member.
Provide specific descriptions:			

(Jan. 2002) Page 2 of2

Probationary Firefighter:	Date:

PERSONAL TRAITS (CONT.)

Decisiveness

| ☐ Consistently demonstrates unsatisfactory performance in promptly making a decision. When questioned on actions or statements, vacillates or changes mind. Unsatisfactory in making decisions at emergency incidents or in stressful situations. Too often fails to think logically or confidently in a timely manner. Noncommittal. Lets or causes others to take charge because of his/her indecisiveness. | ☐ Inconsistent in promptly making a decision and/or standing by the decision. When questioned on actions or statements, occasionally vacillates or changes mind. Occasionally unsatisfactory in making decisions at emergency incidents or in stressful situations. Occasionally does not think or act logically, or confidently in a timely manner. Occasionally is noncommittal. Occasionally lets or causes other to take charge because of his/her indecisiveness. | ☐ Consistently demonstrates a satisfactory ability to make a decision and stands firm on beliefs. However, does recognize the need to correct a wrong decision or incorrect belief and is not afraid to adopt better ideas from other sources. Usually puts being right ahead of being popular or easy. | ☐ Consistently demonstrates an ability to make a decision and stand by it. Able to be decisive even, under difficult conditions. Clearly stands firm in decisions and beliefs. Balances firm resolve and flexibility; recognizes a better idea when it arises and incorporates it into own decisions. Recovers quickly after learning a decision was incorrect. Clearly thinks and acts quickly, logically, confidently. Readily accepts responsibility for decisions; passes credit and shoulders blame. Makes the "hard, right" decision rather than the "easy wrong". |

Provide specific descriptions:

Grooming and Uniform Standards

| ☐ Consistently fails to conform to Department grooming and uniform standards including cleanliness and neatness of uniform and self. | ☐ Sometimes fails to conform to Department grooming and uniform standards including cleanliness and neatness of uniform and self. | ☐ Consistently conforms to Department grooming and uniform standards including cleanliness and neatness of uniform and self. | ☐ Consistently exceeds Department grooming and uniform standards including cleanliness of uniform and self. |

Provide specific descriptions:

Adaptability to Routine

| ☐ Consistently has difficulty adjusting to and/or meeting routine or standard operating procedures or Department rules and regulations. | ☐ On several occasions has difficulty adjusting to and/or meeting routine or standard operating procedures or Department regulations. | ☐ Consistently able to adjust to/meet routine or standard operating procedures and Department rules and regulations. | ☐ Consistently demonstrates an above standard ability to adjust to/meet routine or standard operating procedures and Department rules and regulations. |

Provide specific descriptions:

(Jan. 2002) Page 3 of 3

Probationary Firefighter:	Date:

PERSONAL TRAITS (CONT.)

Verbal Communication

☐ Consistently unable to demonstrate confidence, poise, organization, and skills in communications. Confuses the audience, prompting many questions seeking clarification (e.g., when dealing with the public, presenting proficiency exercises and relating to other members and supervisors). Has made inappropriate comments and gestures regarding human diversity and fails to correct behavior. Fails to convey sincerity or conviction. Often speaks before thinking, hence requires retractions or corrections.	☐ Occasionally unable to demonstrate confidence, poise, organization, and skills in verbal communications. Often confuses the audience, prompting many questions seeking clarification (e.g., when dealing with the public, presenting proficiency exercises, and relating to other members and supervisors). Occasionally fails through comments and gestures to demonstrate respect for human diversity and makes insufficient corrections when advised of their inappropriate behavior. Occasionally fails to convey sincerity or conviction. Sometimes speaks before thinking, hence may occasionally require retractions or corrections.	☐ Consistently demonstrates confidence, poise, organization, and skills in oral communications and with gestures. For instance, when dealing with the public, presenting proficiency exercises, and relating to other members and supervisors. Demonstrates a respect for human diversity. Conveys sincerity and conviction. Generally uses grammatically correct terms and phrases.	☐ Consistently demonstrates an above standard ability in verbal communications when dealing with the public, presenting proficiency exercises, and relating to other members and supervisors. Demonstrates an above average respect for human diversity.
Provide specific descriptions:			

Probationary Firefighter:	Date:

STATION DUTIES

Initiative

☐ Consistently fails to be a self-starter during station duties. Consistently fails to contribute to the problem-solving or decision-making process. Repeatedly needs prompting; does not originate ideas or actions. Fails to seek clarification of obviously unclear guidance. Typically waits for instructions when he/she could act in the absence of guidance. Fails to identify or act on key implied tasks (such as: assists with cooking chores, does extra details, works on station projects). Typically fails to anticipate obvious requirements or contingencies; typically fails to recognize or deal with major problems; sidetracked by irrelevant problems. Easily flustered in a fluid environment. Does not volunteer or does so ineffectively or superficially.	☐ Inconsistent in being self-motivated. Occasionally needs prompting; does not originate ideas or actions. Occasionally fails to contribute to the problem-solving or decision-making process. Occasionally fails to seek clarification of obviously unclear guidance. Tends to wait for instructions when he/she could act in the absence of guidance. Acts on specified tasks only; occasionally fails to identify or act on key implied tasks (such as: assists with cooking chores, does extra details, works on station projects). Occasionally fails to anticipate obvious requirements or contingencies; occasionally fails to recognize or deal with major problems; occasionally side-tracked by an irrelevant problem; occasionally flustered in a fluid environment. Rarely volunteers or does so ineffectively or superficially.	☐ A self-starter with all station duties. Works to correct weaknesses. Asks questions to attempt to clarify instructions or guidance. Usually acts in the absence of guidance. Participates in or contributes to the problem-solving or decision-making process Makes good effort to identify specified and implied tasks (such as: assists with cooking chores, does extra details, works on station projects). Attempts to improvise within supervisor's intent. Volunteers in useful ways.	☐ A self-starter with all station duties. Does markedly more than what is specifically required. Dynamic self-starter; originates ideas or actions. Looks for and displays a willingness to use new methods. Asks pertinent, effective questions that help to clarify instructions or guidance received. Quickly acts in the absence of guidance. Thoroughly seeks out specified and implied tasks (such as: assists with cooking chores, does extra details, works on station projects). Accurately anticipates requirements or contingencies. Improvises well within supervisor's intent; excels in a fluid environment. Readily volunteers; makes a difference for the team.

Provide specific descriptions:

Productivity and Quality of Work

☐ Consistently unable to follow a routine and/or complete a satisfactory amount of work during daily activities. Fails to get a specified or an important implied task done. Does substandard work. Disregards priorities set by supervisor and Department. Too easily stymied by obstacles, difficulties, or hardships. Abuses resources.	☐ Inconsistent in following a routine and/or completing a satisfactory amount of work during daily activities. Needs prompting to get a specified or an important implied task done. Occasionally does substandard work. Occasionally disregards priorities set by supervisor and Department. Occasionally stymied by obstacles, difficulties, or hardships. Inappropriately uses resources.	☐ Consistent in following a daily routine and completing a satisfactory amount of work during daily activities. Gets specified and important implied tasks done. Meets minimum standards. Observes priorities set by supervisor and Department. Copes adequately with obstacles, difficulties, and hardships. Gets things done with available resources.	☐ Demonstrates performance in following a daily routine that is above the standard. Performs an amount of work that is also above the standard and able to complete special assigned duties. Gets specified and implied tasks done right the first time, on time. Exceeds standards. Observes priorities set by supervisor and Department. Impressively overcomes obstacles, difficulties, and hardships. Makes wise use of available resources.

Provide specific descriptions:

Probationary Firefighter:	Date:

EMERGENCY DUTIES & TRAINING

Response Towards Training and Emergency Activities

☐ Consistently demonstrates an unsatisfactory response toward training and emergency activities. Lacks enthusiasm and/or initiative for drills or emergency operations. Makes little or no effort to improve areas needing improvement. Sits back and lets others take the initiative when they are capable of doing so. Does not follow standard operating procedures.	☐ Inconsistent in response toward training and emergency activities. May lack enthusiasm and/or initiative for drills or emergency operations. Makes insufficient effort toward areas needing improvement. Has trouble following or remembering standardized operating procedures.	☐ Consistently demonstrates a satisfactory response toward training and emergency activities. Takes the initiative when appropriate, and is open to suggestions. Understands and follows standardized operating procedures. Works to improve known weak areas without being told.	☐ Consistently demonstrates a response that is positive and above the standard, toward training and emergency duties. Understands and follows standardized operating procedures. Eager to learn all relevant aspects of emergency operations. Enthusiastic and consistently takes the initiative, when appropriate, at emergency incidents. Assertively works to identify and correct areas in need of improvement without being told.

Provide specific descriptions:

Ability to Learn, Retain, and Apply Information

☐ Consistently has difficulty understanding and retaining new material. Does not respond to remedial training. Consistently unable to apply learned information to routine or emergency situations. This inability to apply basic concepts during practical situations may endanger the probationary firefighter or others. Unable to be left unsupervised in situations where sound judgement is required. Unable to function in slightly new situations without strict direction and supervision. Consistently has difficulty when material is reviewed on a spot check basis without the opportunity to consult notes or written material.	☐ On several occasions has difficulty understanding and retaining new material. Requires excessive remedial training. Inconsistent in applying learned information to routine or emergency situations where sound judgement may be required. Has difficulty in functioning in slightly new situations without strict direction and supervision. Sometimes has difficulty when material is reviewed on a spot check basis without the opportunity to consult notes or written material.	☐ Consistently able to understand and retain new material and link it to previously learned related information. Consistently able to apply learned information to routine emergency situations. Able to be left unsupervised in situations where sound judgement may be required. Able to apply normal life experiences to different situations encountered at work. Performs satisfactorily when material is reviewed on a spot check basis.	☐ Overall performance above that required in understanding and retaining new material. Learns quickly, able to combine new material with previous knowledge. Able to generalize and distinguish new information. Demonstrates a superior ability to apply learned information to routine and emergency situations. Able to apply normal life experiences to difficult situations encountered at work. Functions well, without specific instructions to different situations or problems. Demonstrates remarkable knowledge when material is reviewed on a spot check basis.

Provide specific descriptions:

Probationary Firefighter:	Date:

EMERGENCY DUTIES & TRAINING (CONT.)

Effort to Improve

☐ Consistently demonstrates insufficient effort to correct known deficiencies. Often unaware of weaknesses. Even when aware of weaknesses takes little to no action to correct performance.	☐ Often demonstrates insufficient effort in correcting known problem areas. Needs to be prompted to work on areas in need of improvement, often times even when aware of the need.	☐ Consistently demonstrates a "focused, systematic effort" to correct areas in need of improvement. Assertively seeks to learn from experience and discussion with others.	☐ Demonstrates a "zeal" that is above the standard, in seeking to improve his/her performance. Uses every appropriate opportunity to learn from his/her and others' experiences, including mistakes.

Provide specific descriptions:

Follows Directions

☐ Consistently fails to follow directions, orders, and standard operating procedures. Cannot be relied on to perform assigned duties and/or perform as a member of a team.	☐ Inconsistent in following directions, orders, and standard operating procedures. Inconsistent in performing assigned duties and functioning as a member of a team.	☐ Consistently follows directions, orders, and standard operating procedures. Can be relied on to perform assigned and assumed duties and function as part of a team.	☐ Performs above the standard in following directions, orders, and standard operating procedures. Can be relied on to perform assigned duties and function as part of a team even under difficult or very unusual situations.

Provide specific descriptions:

Initiative

☐ Consistently fails to be a self-starter during emergency operations and drills. Frequently must be told to take action that is necessary. Does not move to correct weaknesses.	☐ Inconsistent in being self-starter. Occasionally must be told to take action that is necessary. Does not always move to correct areas needing improvement, and/or do what is assigned to meet job related competencies.	☐ Is a self-starter during training and emergency operations and drills, where appropriate. Works to correct known weaknesses. Consistently seeks to meet or exceed job-related competencies.	☐ Is a self-starter during emergency operations and drills, where appropriate. Consistently demonstrates a "focused, systematic determination" during emergency operations and drills that are above the standard. Does markedly more than what is specifically required.

Provide specific descriptions:

Quality of Work

☐ Consistently demonstrates an inability to perform satisfactorily various operations required by emergency duties and in training.	☐ Inconsistent in performing various operations required by emergency duties and in training.	☐ Consistently demonstrates the ability to perform satisfactorily various operations required by emergency duties and in training.	☐ Consistently demonstrates performance in various operations required by emergency duties and in training that is above the standard.

Provide specific descriptions:

Probationary Firefighter:	Date:

EMERGENCY DUTIES (CONT.)

Strength & Stamina

☐ Consistently demonstrates unsatisfactory strength and/or endurance during required duties and emergency operations or drills. For instance, unable to perform certain ladder evolutions, or other emergency duties such as axe wielding, hoisting lines and equipment. Or unable to perform hard work for other than brief intervals. Tends to quit, complain, or wear out while others keep going. Fades or fails in physical demanding endeavors; malingers. Often needs help from others to get things done. Succumbs to emotional stress or personal fears.	☐ Occasionally fails to demonstrate satisfactory strength and/or endurance during required duties, emergency operations or drills. Occasionally demonstrates a tendency to succumb to emotional stress or personal fears. Occasionally quits, complain, or wears out while others keep going.	☐ Consistently demonstrates satisfactory strength and endurance in the successful completion of all on-duty activities. Adequately copes with emotional stress and personal fears	☐ Consistently demonstrates strength and endurance that is above the standard in the satisfactorily completion of all on-duty activities. Mentally and physically durable even in stressful or strenuous situations. Excels in physically demanding endeavors; helps others succeed. Remains poised and effective despite emotional stress and personal fears.

Provide specific descriptions:

Safety Consciousness

☐ Consistently fails to demonstrate a satisfactory concern for, or awareness of, the safety of themselves or others. OR Their inattention or neglect endangers others. Their inattention or neglect causes injury to themselves or others. OR Not assigned certain functions during emergency operations or drills due to demonstrably unsatisfactory safety consciousness. Requires constant and direct supervision to prevent possible injury to themselves or others. Consistently violates regulations on the use and care of protective clothing and equipment.	☐ Inconsistent in demonstrating a satisfactory concern for, or awareness of, the safety of themselves or others. Does not consistently follow Department regulations on the use and care of protective clothing and equipment. Requires close supervision to prevent possible injury to themselves or others.	☐ Consistently demonstrates a satisfactory concern for, and attention to, safety at emergency incidents and drills. Follows Department regulations on the use and care of protective clothing and equipment, and follows safety procedures. Able to anticipate the potential for injury to themselves or others at emergency incidents and drills.	☐ Demonstrates an awareness of, and concern for safety that is markedly beyond the standard. Consistently follows all relevant regulations and procedures. Able to function with very minimal supervision in potentially unsafe situations. Demonstrates superior ability to anticipate the potential for injury to themselves or others at emergency incidents and drills.

Provide specific descriptions:

Probationary Firefighter:	Date:

EMERGENCY DUTIES & TRAINING (CONT.)

Study Habits – Didactic

☐ Consistently demonstrates an unsatisfactory approach toward learning and retaining written material. Does not utilize on duty study time that is provided at the station. Unable to establish and/or conform to a logical study program.	☐ Occasionally exhibits an unsatisfactory approach toward learning and retaining written material. Sometimes fails to utilize study time properly and conform to a logical study program.	☐ Consistently demonstrates efficient and effective use of study time, and follows prescribed minimum study program. Demonstrates a commitment to a career-long learning process.	☐ Consistently demonstrates performance in establishing and adhering to a prescribed minimum study program that is above the standard. Able to articulate and demonstrate a commitment to a career-long learning life style.

Provide specific descriptions:

Study Habits – Manipulative

☐ Consistently fails to demonstrate proper preparation for practical evolutions. Does not practice without prompting of supervisors or fellow members. Frequently does not make an extra effort to correct known deficiencies.	☐ Occasionally fails to demonstrate proper preparation for practical evolutions. Does not practice without prompting of supervisors or fellow members. Occasionally fails to take extra effort to correct known deficiencies.	☐ Demonstrates a satisfactory adherence to a study routine for practical evolutions. Practices on own initiative, creates and diligently follows an efficient and effective process to correct known deficiencies.	☐ Demonstrates preparation for practical evolutions that is above the standard. Initiates practice for practical evolutions, makes an extra effort to correct known deficiencies.

Provide specific descriptions:

Hose Lays

☐ Consistently fails to demonstrate a satisfactory knowledge of Hose Lays and is unable to perform them.	☐ Occasionally fails to demonstrate a satisfactory knowledge of Hose Lays and the ability to perform them.	☐ Consistently demonstrates a satisfactory knowledge of Hose Lays and is able to perform them.	☐ Demonstrates knowledge and ability with Hose Lays that exceeds the standard.

Provide specific descriptions:

Ropes & Knots

☐ Consistently fails to demonstrate a satisfactory knowledge of knots and/or their practical use in the field.	☐ Occasionally fails to demonstrate a satisfactory knowledge of and use of knots and their practical use in the field.	☐ Consistently demonstrates a satisfactory knowledge of and ability with, knots and their practical use in the field.	☐ Consistently demonstrates knowledge of and ability with, knots and their practical use in the field that is above the standard.

Provide specific descriptions:

(Jan. 2002) Page 9 of 9

Probationary Firefighter:	Date:

EMERGENCY DUTIES & TRAINING (CONT.)

Vehicle Extrication

☐ Consistently fails to demonstrate a satisfactory knowledge of vehicle extrication principles and/or the practical use of the tools needed for extrication.	☐ Inconsistent in demonstrating a satisfactory knowledge of vehicle extrication principles and/or the practical use of the tools needed for extrication	☐ Consistently demonstrates a satisfactory knowledge of vehicle extrication principles and/or the practical use of the tools needed for extrication.	☐ Consistently demonstrates a knowledge of vehicle extrication principles and/or the practical use of the tools needed for extrication that is above the standard

Provide specific descriptions:

Ladders

☐ Consistently fails to demonstrate satisfactory performance with Ground Ladders. Consistently unsatisfactory in such aspects as ladder selection, placement, handling and evolutions under drill and/or emergency conditions.	☐ Inconsistent in demonstrating satisfactory performance with Ground Ladders. Occasionally fails in such aspects as ladder selection, placement, handling and evolutions under drill and/or emergency conditions.	☐ Consistently performs in a satisfactory manner with Ground Ladders. Demonstrates satisfactory performance in all aspects of ladder handling such as selection and placement under drill and emergency conditions.	☐ Consistently demonstrates above standard performance in both the knowledge and use of all Ground Ladders, even under stressful drill and emergency conditions.

Provide specific descriptions:

Ventilation (Principles, Concepts, Anatomy of an Opening)

☐ Consistently fails to demonstrate satisfactory performance with ventilation principles, concepts, and anatomy of an opening. Consistently unsatisfactory in aspects of building construction and how ventilation aspects apply.	☐ Inconsistent in demonstrating satisfactory performance with ventilation principles, concepts, and anatomy of an opening. Occasionally fails to recognize aspects of building construction and how ventilation aspects apply.	☐ Consistently performs in a satisfactory manner with ventilation principles, concepts, and anatomy of openings. Demonstrates satisfactory performance in all aspects of building construction and how ventilation aspects apply.	☐ Consistently demonstrates above standard performance in ventilation principles, concepts, anatomy of openings, building construction and how ventilation aspects apply; even under stressful drill and emergency conditions.

Provide specific descriptions:

Forcible Entry (Principles and use of tools)

☐ Consistently fails to demonstrate a satisfactory knowledge of forcible entry principles and/or the practical use of the tools needed for forcible entry.	☐ Inconsistent in demonstrating a satisfactory knowledge of forcible entry principles and/or the practical use of the tools needed for forcible entry.	☐ Consistently demonstrates a satisfactory knowledge of forcible entry principles and/or the practical use of the tools needed for extrication	☐ Consistently demonstrates a knowledge of forcible entry principles and/or the practical use of the tools needed for forcible entry that is above the standard

Provide specific descriptions:

Probationary Firefighter:	Date:

EMERGENCY DUTIES & TRAINING (CONT.)

Tools & Equipment

☐ Consistently demonstrates an unsatisfactory knowledge and/ or use of Tools and Equipment. Consistently poor in verbal presentation during drills or questioning or during practical application of tools and equipment to various tasks. Performs Primarily at a General Knowledge Level.	☐ Inconsistent in demonstrating satisfactory knowledge and use of Tools and Equipment. Sometimes fails to satisfactorily answer questions, or follow proper procedures during "hands on" drills. Performs primarily at a Working Knowledge level.	☐ Consistently demonstrates satisfactory knowledge and use of all aspects of Tools and Equipment. Able to satisfactorily answer questions and present drills, in addition to applying academic knowledge to practical situations. Consistently performs at a Qualified Knowledge Level when using Tools and Equipment in "real life" situations.	☐ Consistently demonstrates superior performance in both knowledge and use of all aspects of Tools and Equipment. Able to apply skills under a wide variety of circumstances and consistently performs at a Qualified Knowledge Level. Has a mechanical aptitude that allows them to maximize the potential of assigned Tools and Equipment.

Provide specific descriptions:

EMS – Performs initial assessment (60 seconds) and intervenes immediately as indicated. Initial assessment: Environment, ABC's, LOC, Skin Vitals, Chief Complaint

☐ Omits portions of initial assessment, or fails to intervene when a problem is noted.	☐ Performs a complete initial assessment, but is either very slow in assessing and/or intervening or is disorganized.	☐ Performs a complete and fairly organized initial assessment in a reasonable amount of time. Recognizes critical vs. non-critical patient, i.e., opens airway; intervenes appropriately.	☐ Demonstrates above average skills in performing a complete and organized initial assessment in a timely manner. Intervenes rapidly, initiates CPR, etc.

Provide specific descriptions:

EMS – Obtains relevant and accurate patient history, chief complaint, medications and allergies in a systematic manner (Focused history and detailed physical examination).

☐ Totally disorganized patient assessment. Does not get pertinent information. Is not complete. Report writer frequently needs to request additional information.	☐ Obtains an adequate patient assessment, but is either very slow in assessment and/or is disorganized.	☐ In a reasonable amount of time, obtains an adequate patient history, chief complaint, medications and allergies in a fairly organized manner. (Adequate is defined as acceptable, but not remarkable).	☐ Able to gather information efficiently even in difficult situations (i.e., emotionally charged situation). Organized and timely.

Provide specific descriptions:

(Jan. 2002) Page 11 of11

Probationary Firefighter:	Date:

EMERGENCY DUTIES & TRAINING (CONT.)

EMS – Recognizes patients that need further medical attention, determines appropriate mode of transport (ALS or BLS ambulance, private car, etc.) and transports at appropriate point in run.

☐ Frequently fails to recognize patients requiring further medical attention and/or does not use good judgement in determining appropriate mode of transportation. Frequently fails to recognize transport needs. Makes the same or similar mistake on subsequent runs.	☐ Usually recognizes patients needing further medical attention but does not consistently choose correct mode of transport. Occasionally fails to recognize the transport needs. Does not make the same or similar mistakes on subsequent runs.	☐ Consistently uses good judgement in determining patients needing further medical attention. Errs on side of patient. Chooses appropriate mode of transportation. Determines appropriate time in run to transport: a. "scoop & run" b. deterioration of patient status c. patients requiring definitive care at scene d. patients not responding to field tx	☐ Uses excellent judgement regarding medical attention/ transportation. Shows strong ability to identify appropriate transportation needs in even the most difficult situations.

Provide specific descriptions:

EMS – Obtains vital signs quickly and accurately when indicated.

☐ Frequently does not take vital signs at appropriate time, or frequently has problems with procedure, or vital signs obtained are inaccurate.	☐ Occasionally vital signs are not put in proper priority. Obtains vital signs accurately, but takes too long to perform procedure.	☐ Usually takes vital signs at the appropriate times. Obtained in a reasonable amount of time; accurate.	☐ Shows a strong ability to correlate vital signs with patient's condition. Consistently accurate and timely.

Provide specific descriptions:

EMS – Interprets assessment information correctly and takes appropriate action.

☐ Frequently is unable to interpret assessment information correctly. Demonstrates weak knowledge base, or suggests treatments that would have an adverse effect on the patient.	☐ Interprets assessment information correctly, but is hesitant about action to take.	☐ Correlates information gained from patient assessment with knowledge base gained from training and suggests treatment appropriate to situation.	☐ Shows strong ability to interpret assessment information and take appropriate action.

Provide specific descriptions:

EMS – CPR

☐ Demonstrates poor technique, i.e., hand position, airway management, comp/vent ratio.	☐ Performs CPR in a consistent manner, but fails to correct other personnel when performance is inadequate.	☐ Performs CPR according to AHA standards.	☐ Demonstrates skills and knowledge at instructor level.

Provide specific descriptions:

(Jan. 2002) Page 12 of 12

Probationary Firefighter:	Date:

EMERGENCY DUTIES & TRAINING (CONT.)

EMS – Airway Control

☐ Frequently fails to open and maintain an airway. Uses or does not recognize poor technique.	☐ Is inconsistent. On occasion, adequately manages airway and occasionally does not.	☐ Consistently opens and maintains an adequate airway using appropriate technique. Treats it appropriately and as highest priority.	☐ Same as previous rating with remarkable knowledge of procedure.

Provide specific descriptions:

EMS – Bandaging and Splinting

☐ Ineffective technique or treatment causing potential harm to patient. Failure to initiate any treatment.	☐ Aware of need for bandaging/ splinting, but needs some direction (slow but adequate).	☐ Applies appropriate and adequate dressings/ splints in a timely manner.	☐ Able to dress difficult/ unique injuries well. Very neat.

Provide specific descriptions:

EMS – Oxygen Administration (cannula, mask, PPV)

☐ Fails to administer O2 when indicated, **or** fails to use equipment properly.	☐ Inconsistent – Occasionally fails to initiate O2 therapy when indicated. Uses equipment properly.	☐ Administers O2 when necessary. Selects correct method/ rate of administration. Uses equipment proficiently.	☐ Same as previous rating, with remarkable knowledge of O2 administration.

Provide specific descriptions:

EMS – Spinal Immobilization

☐ Fails to recognize obvious back/ neck injuries. Fails to initiate spinal precautions in obvious situations. Incomplete or incorrect procedure.	☐ Recognizes obvious problems but may fail to recognize potential problems and is slow to initiate proper treatment. Incomplete treatment, i.e., C-collar only/ backboard only. Inconsistent performance of procedure.	☐ Uses full spine precautions whenever indicated. Recognizes obvious and potential problems. Uses complete and correct procedure.	☐ Consistently above average application and anticipation.

Provide specific descriptions:

Appendices

Probationary Firefighter:	Date:

FIRE PREVENTION / PUBLIC RELATIONS

Response Towards Fire Prevention Activities

☐ Consistently demonstrates an unsatisfactory response toward Fire Prevention. Lacks enthusiasm and/or initiative for Fire Prevention activities such as conducting inspections, notice preparations and record keeping. Sits back and lets others take the initiative when they are capable of doing so. Makes little or no effort to improve areas needing improvement.	☐ Inconsistent in response toward Fire Prevention activities. May lack enthusiasm and/or initiative for Fire Prevention activities. Makes insufficient effort toward areas needing improvement.	☐ Consistently demonstrates a satisfactory response toward Fire Prevention activities. Shows enthusiasm, takes the initiative when appropriate, and open to suggestions. Utilizes Fire Inspections for pre-fire, pre-collapse and other rescue planning. Works to improve known weak areas without being told.	☐ Consistently demonstrates a response towards Fire Prevention activities that is positive, and above the standard. Eager to learn all relevant aspects of Fire Prevention activities. Consistently takes the initiative when appropriate. Assertively works to identify and correct areas in need of improvement without being told.

Provide specific descriptions:

Code Enforcement

☐ Consistently demonstrates significant deficiencies in both knowledge and enforcement of the Fire Code.	☐ Sometimes fails to meet standards in knowledge and enforcement of the Fire Code.	☐ Demonstrates good working knowledge of the Fire Code and enforces its provisions.	☐ Demonstrates knowledge of the Fire Code and enforcement of its provisions that is above the standard.

Provide specific descriptions:

Public Relations

☐ Consistently demonstrates unsatisfactory behavior toward the public. Does not act to further a favorable response to Department activities. Exhibits improper behavior when dealing with the public.	☐ Sometimes fails to demonstrate satisfactory behavior toward the public. Inconsistent in acting to further a favorable public response to Department activities.	☐ Consistently demonstrates satisfactory behavior toward the public, even under difficult conditions. Seeks to further Fire Prevention and Fire Safety themes when dealing with the public. Represents the Department satisfactorily in all on-duty contacts. Is courteous when inspecting; takes appropriate action with complaints and requests for service.	☐ Consistently demonstrates behavior toward the public that is above the standard. Acts to further Fire Prevention and Fire Safety themes when dealing with the public. Deals courteously with all members of the public, even under very trying conditions. Is courteous when inspecting; takes appropriate action and complaints and requests for service. Demonstrates exceptional "vigor" in all on-duty contacts with the public.

Provide specific descriptions:

(Jan. 2002) Page 14 of 14

276

Probationary Firefighter:	Date:

FIRE PREVENTION / PUBLIC RELATIONS (CONT.)

Quality of Work

☐ Consistently demonstrates unsatisfactory performance in all or some aspects of Fire Prevention activities.	☐ Occasionally fails to perform satisfactorily in some aspects of Fire Prevention activities.	☐ Consistently demonstrates satisfactory performance in all aspects of Fire Prevention activities.	☐ Consistently demonstrates a performance in all aspects of Fire Prevention activities that is above the standard.

Provide specific descriptions:

Record Keeping

☐ Consistently fails to satisfactorily understand and use the Department's record keeping system. Makes inaccurate and/or illegible entries, omits entries, fails to follow prescribed procedures, or takes excessive time to accomplish assigned tasks.	☐ Inconsistent in satisfactorily understanding and using the Department's record keeping system. Occasionally makes inaccurate and/or illegible entries, omits entries, and fails to follow prescribed procedures, or takes excessive time and material to accomplish assigned tasks.	☐ Consistently demonstrates satisfactory performance of duties. Records are accurate and complete, proper procedures followed, and tasks completed in a timely manner.	☐ Consistently demonstrates a performance of duties that is above the standard. Performs assigned and extra duties effectively and efficiently.

Provide specific descriptions:

Hazard Recognition

☐ Consistently fails to recognize potential or existing fires and life safety hazards. Consistent with level of training.	☐ Occasionally fails to recognize potential or existing fire or life safety hazards. Consistent with level of training	☐ Consistently recognizes potential and existing fire and life safety hazards. Consistent with level of training.	☐ Consistently above the standard in recognition of potential and existing fire and life safety hazards. Consistent with level of training.

Provide specific descriptions:

Notice Preparation

☐ Consistently fails to properly utilize and prepare Fire Prevention or other forms.	☐ Inconsistent in properly utilizing and preparing Fire Prevention or other forms.	☐ Consistently utilizes and prepares Fire Prevention forms in a proper fashion.	☐ Consistently demonstrates a utilization and preparation of Fire Prevention forms that is above the standard.

Provide specific descriptions:

| Probationary Firefighter: | | Date: | |

Comments:

Plan for improvement:

Probationary Firefighter Signature   Date

Captain Signature   Date

Captain Signature   Date

Battalion Chief Signature   Date

Assistant Chief Signature   Date

Director of Training   Date

Smoke Your Firefighter Written Exam
Maximize Your Score While Your Competition Struggles

Most fire departments give a written examination as the first part of the testing process. Since there are usually an enormous number of candidates who apply for the position, difficult written examinations are an effective way to narrow the field.

There is no excuse for a firefighter candidate to fail a written exam. Those who are committed to their goal will plan and prepare for the written exam to improve their chances of success.

While there are other self-help books devoted to helping aspiring firefighters, they only provide questions and answers. These books are useless if you do not understand how to work out the problems.

Smoke Your Firefighter Written Exam goes beyond providing sample questions and answers. It teaches the reader the basic rules and principles behind the questions; in other words, how to solve each complex problem. Each section begins with an overview of how to solve the problem. The reasoning behind the correct answer is presented in clear, easy-to-understand language.

The reader will become proficient in:
- Mathematics (word problems, geometry, addition, subtraction, multiplication)
- Mechanical aptitude (gears, levers, pulleys, incline planes, screws)
- Language (reading, spelling, grammar, vocabulary)
- Perceptual ability
- Spatial relations
- Matching parts and figures
- Map reading

Two ways to order:
1. Order online at www.aspiringfirefighters.com.
2. Call the toll-free order line at: 800-215-9555.

Become a CPR and First Aid Instructor

Enhance your resume.

Develop your public speaking skills.

Offer a vital service to the community.

Perfect your CPR and First Aid techniques.

Support your firefighter job search by turning your EMT knowledge and skills into a viable business. With EMS Safety, you can train to be a CPR, First Aid and AED Instructor. Learn to teach these life-saving skills to businesses, schools and the general public.

(800) 215-9555
www.emssafety.com